ELEMENTS OF
SOLID STATE THEORY

ELEMENTS OF
SOLID STATE THEORY

BY
GREGORY H. WANNIER
Bell Telephone Laboratories

GREENWOOD PRESS, PUBLISHERS
NEW YORK

Originally published in 1959
by the Syndics of the Cambridge University Press

Reprinted by permission
of Cambridge University Press

First Greenwood Reprinting 1970

Library of Congress Catalogue Card Number 79-99504

SBN 8371-3030-1

PRINTED IN UNITED STATES OF AMERICA

CONTENTS

PREFACE

This book represents an attempt to provide a coherent and brief outline of the theory behind present-day solid state physics. It is aimed at advanced students or scientists with good all-round knowledge, but does not assume previous acquaintance with this specialty. An effort was made to produce a book which can be read from cover to cover and which reveals, if read in this way, some of the main ideas specific to solid state physics. In order to achieve this, severe limitations in scope had to be accepted. Some topics of current interest are missing altogether, and others will be found in skeletonized form only. To minimize this latter inconvenience a bibliography of comprehensive articles is appended to each chapter through which the topics discussed in the book may be opened up fully. Also the problems at the end of the chapters will help to round out any particular discussion which may appear too terse to the reader.

It is a pleasure to acknowledge the help and encouragement received from many sources. Students at Bell Laboratories, Princeton University and the University of Geneva were good enough to act as audiences, and through their questions, helped to remove many obscurities. Among colleagues I am particularly indebted to A. N. Holden and E. Vogt. Alan Holden's parallel endeavours helped to resolve some difficult questions of presentation, particularly in the theory of bonding. Erich Vogt read and criticized in detail the entire manuscript. Further thanks are due to E. N. Adams, M. H. Cohen, J. K. Galt and J. C. Phillips who read selected chapters and in some cases prevented errors from going to print. Acknowledgement is given in the text to authors who permitted reproduction of figures or tables. Many people helped me in the discussion of special points, in particular P. W. Anderson, C. J. Ballhausen, C. Herring, H. Suhl and E. A. Wood. Finally, I want to thank the staffs of Bell Laboratories and New Providence Typing Service for their unquestioning helpfulness, and my wife Carol Wannier for reading the proofs. G.H.W.

MURRAY HILL
December 1958

ERRATA

p. 14. The text implies incorrectly that there is no three-dimensional equivalent of Theorem 1. Actually the equivalent theorem reads:

THEOREM 1 *a*. *A Bravais lattice has inversion symmetry about every lattice point and about every midpoint between two lattice points.*

Actual crystals may or may not have this symmetry.

p. 72. In the third equation down on the left the function should be $\sinh \xi$.

p. 80. The T^4-term in (3.62 *b*) is the Stefan–Boltzmann law as applied to the sound field. Thus it yields no information about the atoms or molecules as is evident from p. 68 and (3.59). Only the breakdown of this law at higher temperatures involves the atoms.

p. 90. The right-hand side of (4.5) should have a minus sign.

p. 93. According to the convention followed in this book the variable entering into (4.17 *a*) is $U - MH$, not U.

p. 97. Factor on the right of (4.22) should be $\sqrt{3}$.

p. 101. Equation (4.32). Same observation as for p. 93.

p. 132. Equation (5.6). This equation is later referred to as Floquet's theorem.

pp. 134–5. The proof given that E varies monotonely with $\cos 2\pi kd$ in every band is defective. A good proof is in E. C. Titchmarsh, *Eigenfunction Expansions*, Part II (Oxford University Press, 1958). It fills the first half of § 21.4, p. 292.

p. 144. The ordinates on figs. 5.6 and 5.7 should be labelled $8d^2 mW/h^2$.

p. 156. The reference to (5.68) in the middle of the page should be to (5.70).

p. 163. Part (*d*) of problem 5 should be a separate problem. It deals with an energy band function of the form

$$W = W_0 \cos \alpha k_x \cos \alpha k_y.$$

p. 200. The right-hand side of (6.88) should have a plus sign.

p. 203. The note in problem 9 makes its solution harder. Use instead the conventional gauge and study the change in time of the operators (5.39).

CHAPTER I

GEOMETRY OF THE CRYSTALLINE STATE

Order is one of the main concepts which permits an approach to the solid state, in the same sense that disorder is a powerful theoretical weapon for the study of gases. Some properties of solids are a direct consequence of this state of order (for instance, selective reflection of X-rays) while in many other phenomena order is in fact not physically fundamental, but its presence simplifies the theoretical basis sufficiently to make a deductive procedure possible (for instance, the frictional processes producing Ohm's law). Thus, it is useful to start a study of the solid state with a study of its elements of order. In making use of this notion one realizes soon that the idea of geometrical order ramifies into a large number of special types of order, each one with its individual character. Correspondingly, while there is only one type of gaseous phase, there are hundreds of different crystalline solids. We shall discuss in this section some of the simpler notions connected with the structure of crystals. An exhaustive treatment is clearly not wanted here, for the reasons just stated. Indeed, the subject is sufficiently vast to form a separate scientific discipline, called mathematical crystallography.

A rigid body is called *symmetric* if it remains physically identical to itself under a *translation, rotation,* or *reflection,* or a combination of these operations. The operations are called the *symmetry operations* for the body. A succession of two symmetry operations is also a symmetry operation, and every such operation has an inverse which is also a symmetry operation. Operations for which these two postulates are satisfied are said to be *elements* of a *group.* There are many different kinds of crystalline symmetry groups or *space groups*; for any given rigid body its space group is specific, and forms in some way a description of it, though admittedly not a complete one. All the space groups are subgroups of a group which contains them all; namely, the group of all translations, rotations and reflections in solid geometry. The general element of this group can be represented by its associated transformation in rectangular Cartesian co-ordinates; this has the form

$$\begin{pmatrix} x_1' \\ x_2' \\ x_3' \end{pmatrix} = \begin{pmatrix} \alpha_{11} \alpha_{12} \alpha_{13} \\ \alpha_{21} \alpha_{22} \alpha_{23} \\ \alpha_{31} \alpha_{32} \alpha_{33} \end{pmatrix} \begin{pmatrix} x_1 \\ x_2 \\ x_3 \end{pmatrix} + \begin{pmatrix} a_1 \\ a_2 \\ a_3 \end{pmatrix}, \qquad (1.1)$$

with the restriction on the matrix that the sum of the squares of the elements in any line or column be equal to 1, and that the sum of the products two by two of two parallel lines or columns be zero. It follows from this that the determinant of the α's is ± 1, the $-$ sign belonging to operations involving an odd number of reflections. Sometimes it is convenient to consider subgroups of this general group. Thus, suppression of the operations of determinant -1 leaves the operations which can be physically carried out on a rigid body by moving it. Within this subgroup, a more restricted subgroup is selected by considering only matrices of the form

$$\begin{pmatrix} 1 & 0 & 0 \\ 0 & 1 & 0 \\ 0 & 0 & 1 \end{pmatrix},$$

which yields the group of all parallel translations. The *crystal translation group* is an even more restricted subgroup of this, namely it consists of the parallel translations which are symmetry operations for a given crystal when assumed of infinite extent. This group will play an important role in this outline. Another type of specialization is employed in classical crystallography, which concerns itself particularly with symmetry groups for crystals in which the vector $\begin{pmatrix} a_1 \\ a_2 \\ a_3 \end{pmatrix}$ in (1.1) is suppressed, that is, groups containing reflections and rotations only. Such symmetry groups are called *point groups*. It so happens that the translational symmetries of actual crystals have such a small period that they can be detected only by special techniques, such as X-rays. The immediately apparent elements of symmetry of a crystal therefore form a point group. The student should be warned, however, that the point group of a crystal may contain elements which are absent from its true space group. Quartz, for instance, forms the well-known six-sided prisms; this implies for all its gross properties that a rotation by 120° about its axis is a symmetry operation for the crystal. Detailed examination of the structure of quartz shows, however, that the atoms form a spiral staircase arrangement about the axis, and that the rotation by 120° is a symmetry operation only when coupled with a translation along the axis. This translation is of course too small to be observed by macroscopic methods.

Since there are many different crystalline space groups possible, a geometrical description is always, in some way, a catalogue of types. As the number of types is quite large, people have looked for and found simpler methods of classification. The simplification of

Fig. 1.1. Analysis of translational periodicity in one dimension.

classical crystallography, namely to consider point groups only, is not a good starting point for the physicist, because this procedure suppresses the translational symmetry, which is the element most fruitful in consequences. It is preferable for the purpose of deduction to start with this very element. It is in fact sufficient to assume that crystals have translational symmetry in three dimensions in order to deduce from this all the combinations of symmetry elements which are possible. We shall follow this procedure a certain way and indicate the general ideas and results beyond.

To start with, we assume that crystals have a translational period which is capable of being continued indefinitely in a certain direction. In other words, if we consider a definite place in the crystalline arrangement and put a dot there, similar dots occur at equidistant intervals along some straight line. The notion of structure implies that there must be a minimum distance of this type. This implies immediately that all other distances must be integral multiples of this one, for if there were a period which is a fractional multiple, then there would be a point A in fig. 1.1 lying in between two points of the regular sequence; this would define a smaller period, contrary to hypothesis. We shall introduce a vector **a** to correspond to the elementary translation along this line. In a similar way, we can pick a second vector **b** and a third **c** to form a lattice structure as shown in fig. 1.2. If we pick **a**, **b**, **c** in the following special way: |**a**| shall be the shortest period in the lattice (or one of several equal shortest ones), **b** shall be the shortest not parallel to **a**, and **c** the

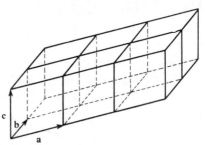

Fig. 1.2. Translational periodicity in three dimensions with basis vectors **a**, **b**, **c**.

shortest not coplanar to **a** and **b**, then the following theorem can be proved:

Any period **r** *of the lattice is a combination of integral multiples of* **a**, **b**, *and* **c**

$$\mathbf{r} = l\mathbf{a} + m\mathbf{b} + n\mathbf{c}; \quad l, m, n \text{ integer.} \tag{1.2}$$

We prove this by showing that if fractional values of l, m, n occur, the method of picking **a**, **b**, **c** must have been violated. Indeed, consider such a point P of vector **r**, at first not assumed in an (**a**, **b**) plane (fig. 1.3). Its perpendicular projection on the nearest (**a**, **b**) plane shall be Q. The perpendicular projection of Q on the nearest line parallel to **a** shall be R and the nearest lattice point to R on that line shall be S. Then

$$(\overline{PS})^2 = (\overline{PQ})^2 + (\overline{QR})^2 + (\overline{RS})^2 \leqq (\tfrac{1}{2}\mathbf{c})^2 + (\tfrac{1}{2}\mathbf{b})^2 + (\tfrac{1}{2}\mathbf{a})^2 \leqq \tfrac{3}{4}\mathbf{c}^2 < \mathbf{c}^2.$$

This result contradicts the way **c** was chosen. If, on the other hand, P is in an (**a**, **b**) plane then P and Q are identical and

$$(\overline{PS})^2 = (\overline{QS})^2 \leqq (\tfrac{1}{2}\mathbf{b})^2 + (\tfrac{1}{2}\mathbf{a})^2 \leqq \tfrac{1}{2}\mathbf{b}^2 < \mathbf{b}^2.$$

This contradicts the choice of **b** unless P is along an **a**-axis. Finally, this last possibility contradicts the choice of **a**.

The lattice exhibited in fig. 1.2 is called the *Bravais lattice* of the crystal to which it belongs[†] and the vectors **r** in (1.2) are its *lattice vectors*. The Bravais lattice has a number of useful properties. One of them is that the symmetry of Bravais lattices is simpler than that of crystals generally. Hence Bravais lattices can be used for a preliminary symmetry classification of crystals. This type of classification will be outlined below.

Let us return for a moment to the original crystal from which the Bravais lattice of fig. 1.2 was derived. This crystal is entirely

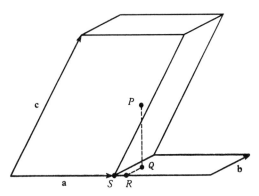

Fig. 1.3. Analysis of translational periodicity in three dimensions.

described if the content of one parallelepiped generated by **a**, **b**, and **c** is known. For the entire remainder of the crystal is generated by periodic repetition of this unit. The unit is called a *primitive cell* of the crystal. The volume of the cell is clearly $\mathbf{a} \cdot (\mathbf{b} \times \mathbf{c})$. Three vectors which generate a primitive cell are said to form a set of *basis vectors*. The question whether three vectors

$$\left.\begin{aligned}
\mathbf{u} &= l\,\mathbf{a} + m\,\mathbf{b} + n\,\mathbf{c}, \\
\mathbf{v} &= l'\mathbf{a} + m'\mathbf{b} + n'\mathbf{c}, \\
\mathbf{w} &= l''\mathbf{a} + m''\mathbf{b} + n''\mathbf{c}, \\
l, m, l' &\ldots \text{etc., integers,}
\end{aligned}\right\} \tag{1.3}$$

are a set of basis vectors is resolved in the following way. We must be able to use them to construct the earlier basis vectors **a**, **b**, **c**; that is, the equation

$$\mathbf{a} = x\,\mathbf{u} + y\,\mathbf{v} + z\,\mathbf{w}$$

[†] M. A. Bravais, *J. Éc. polyt. Paris*, **19** (1850), 1. Available in English translation as a Memoir of the Crystallographic Society of America (1949).

must be solvable in integers x, y, z, and similarly for **b** and **c**. Now this equation when written out says

$$lx + l'y + l''z = 1,$$

$$mx + m'y + m''z = 0,$$

$$nx + n'y + n''z = 0,$$

which solves as

$$x = \frac{\begin{vmatrix} 1 & l' & l'' \\ 0 & m' & m'' \\ 0 & n' & n'' \end{vmatrix}}{\begin{vmatrix} l & l' & l'' \\ m & m' & m'' \\ n & n' & n'' \end{vmatrix}}, \quad y = \ldots.$$

All these numbers will be integers if

$$\Delta = \begin{vmatrix} l & l' & l'' \\ m & m' & m'' \\ n & n' & n'' \end{vmatrix} = \pm 1. \tag{1.4}$$

Hence this condition is sufficient. To show that it is also necessary, consider the primitive cell generated by these vectors; its volume ω^+ is equal to

$$\omega^+ = (l\mathbf{a} + m\mathbf{b} + n\mathbf{c}) \cdot [(l'\mathbf{a} + m'\mathbf{b} + n'\mathbf{c}) \times (l''\mathbf{a} + m''\mathbf{b} + n''\mathbf{c})]$$

$$= \begin{vmatrix} l & m & n \\ l' & m' & n' \\ l'' & m'' & n'' \end{vmatrix} \mathbf{a} \cdot (\mathbf{b} \times \mathbf{c}) = \Delta \mathbf{a} \cdot (\mathbf{b} \times \mathbf{c}).$$

If Δ is not ± 1, this volume is an integral multiple of the original primitive cell volume $\omega = \mathbf{a} \cdot (\mathbf{b} \times \mathbf{c})$. Hence, if the new vectors are used to generate a Bravais lattice, the density of lattice points will be lower and some points will have been missed.

It is important to realize that the criterion just derived for choosing basis vectors permits us to use for one of them (say, the first) any lattice vector whatever, as long as it is the shortest possible in its own direction. To prove this we observe that this means choosing l, m, n in (1.3) with no common divisor except 1, but otherwise arbitrary. Now by the principles of elementary number theory we can always find three numbers L, M, N such that

$$Ll + Mm + Nn = 1.$$

If we can further insure that there are integers l', m', n', l'', m'', n'' such that

$$L = m'n'' - n'm'',$$

$$M = n'l'' - l'n'',$$

$$N = l'm'' - m'l'',$$

our determinant criterion (1.4) will have been satisfied. This can be done. Let d be the greatest common divisor of L and M. We then satisfy the first two equations by setting

$$n'' = 0, \quad n' = -d, \quad m'' = +L/d, \quad l'' = -M/d,$$

whereupon the last equation becomes

$$N = l'\frac{L}{d} + m'\frac{M}{d}.$$

By the principle previously stated integers l', m' satisfying this equation can always be found because L/d and M/d are prime to each other.

It is important for some applications that one vector of a basic set can be chosen arbitrarily. In a similar way, it can be shown that the first two vectors can be chosen so as to generate an arbitrary lattice plane. This will be left as an exercise for the reader.

MATHEMATICAL THEORY OF PLANE BRAVAIS NETS

We are ready now to discuss classification of Bravais lattices according to their symmetry. The method we shall follow is to give a logically coherent discussion of plane Bravais nets. It so happens that this yields essentially all possible elements of symmetry of space lattices. The only things which remain open are the various possible ways in which these Bravais nets might be combined to form space lattices. We shall be satisfied here to cover these latter combinations by a more descriptive discussion with a rough outline of the logical connection.

Fig. 1.4 shows a Bravais net whose possible symmetries we shall now logically deduce. The two elements of symmetry which are sufficient for classification are the line of symmetry (fig. 1.5)

Fig. 1.4. Picture of a Bravais net in two dimensions.

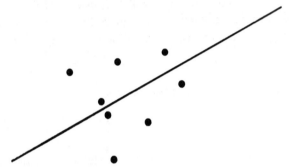

Fig. 1.5. Line of symmetry of a plane figure.

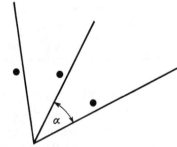

Fig. 1.6. Axis of symmetry of a plane figure with an angle of rotation α.

in the plane and an axis of rotational symmetry at right angles to the plane (fig. 1.6). It is clear that this angle must always be a sub-multiple of 2π. If the angle is $2\pi/n$ the axis is referred to as an *n-fold axis*. From this definition and inspection of fig. 1.4 we derive immediately

THEOREM 1. *There is a twofold axis going through every lattice point of a Bravais net and every midpoint between two lattice points.*

The theorem thus indicates an element of symmetry for Bravais nets which is automatic. However, it is clear that it may be missing in the two-dimensional 'crystal' from which the net is derived. A little reflection shows also that this element of symmetry is not necessarily present in a three-dimensional Bravais lattice.

A further restriction on possible types of symmetry is achieved by

THEOREM 2. *If a Bravais net has a line of symmetry, it has a second line of symmetry at right angles to the first.*

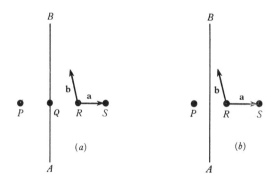

Fig. 1.7. Illustration of Theorem 2 on Bravais nets: pairwise occurrence of lines of symmetry.

To prove this, let AB be such a line of symmetry. Since we do not know in advance whether it does or does not contain a lattice point, we make two drawings, fig. 1.7 (*a*), and fig. 1.7 (*b*), covering either alternative. Now the direction at right angles to AB must be that of a lattice vector because the symmetry operation itself will generate vectors in that direction. It follows from our previous work that the shortest distance in that direction can be taken as one of the basis vectors of the net. A point P lying closest to $\overline{AB}$ will therefore generate a row of equidistant points $PQRS\ldots$ or PRS, according to whether there is or is not a point Q on the line of symmetry. However, the arrangement of fig. 1.7 (*b*) is just as definite as that of fig. 1.7 (*a*) because R must be the mirror image of P and hence the line AB must bisect PR. Since the vector RS may be taken as one of the basis vectors $\mathbf{a}$, the remainder of the net is

2   WST

generated by associating with **a** a second vector **b** and by laying down rows parallel to the row PS with displacements **b**, 2**b**, 3**b**, etc. However, if AB is to be a line of symmetry, the vector **b** cannot be arbitrary. Each time it must produce either the arrangement of fig. 1.7 (*a*) or of fig. 1.7 (*b*) If it repeats the starting arrangement then **b** may be taken at right angles to **a** and we get a rectangular net (fig. 1.8). If it produces the other arrangement the rows alternate and a rhombic net (fig. 1.9) is generated. Both these nets admit any of the horizontal rows as lines of symmetry; this proves the theorem.

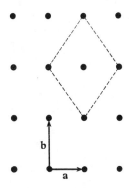

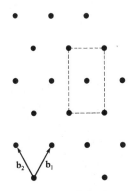

Fig. 1.8. Rectangular Bravais net with basis vectors; the dotted outline shows that it can be considered centred rhombic.

Fig. 1.9. Rhombic Bravais net with basis vectors; the dotted outline shows that it can be considered centred rectangular.

We see that the proof of Theorem 2 has yielded additional information, namely, the two Bravais nets which have the symmetry elements postulated in Theorem 2. They are the *rectangular net* with basis vectors **a** and **b** at right angles (fig. 1.8) and the *rhombic net* with two basis vectors b_1 and b_2 of equal length (fig. 1.9). Since their basic symmetry is the same, their relationship is close. Indeed, one can be generated from the other by placing an extra point in in the centre of the basic parallelogram of the other; this is shown in the two figures. Hence the rhombic net is *face-centred* rectangular and vice versa. Such a centring operation does only

(*a*) Subdivide the previously basic translational period,

(*b*) Intercalate lines of symmetry parallel to already existing lines of symmetry.

Neither one of these two operations is observable by the methods of macroscopic crystallography which only identifies the direction of planes and measures angles. Hence, in flat-land, these two nets would belong to the same 'crystal system'.

The possibility of more than two directions for the lines of symmetry is connected with the appearance of rotational symmetry through

THEOREM 3. *If a figure has two lines of symmetry making an angle α, it has also rotational symmetry about their intersection with angle 2α.*

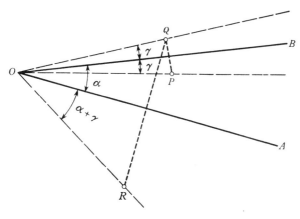

Fig. 1.10. Illustration of Theorem 3 on Bravais nets: at the intersection of two lines of symmetry *A* and *B* there is an axis of symmetry.

To prove this we consider the result of two successive reflections at each line. It is known that this is equivalent to a rotation; the axis of rotation must be the intersection of the two symmetry planes perpendicular to the figure and containing the lines of symmetry. The angle is easily read off from fig. 1.10: *OA* and *OB* are the two lines of symmetry, and *P* is an arbitrary point whose intermediate image is *Q*, and the final image *R*. The statement of the theorem is verified from the angles marked.

It is clear that this theorem applies to anything since no special property of plane Bravais nets was involved. If α = 90° we simply prove again Theorem 1 for things obeying Theorem 2. For plane

Bravais nets this is no news, but it will be of importance in three dimensions. Theorem 3 restricts, by the way, the angle α automatically to a submultiple of π since 2α must be a submultiple of 2π. The main use we shall make of Theorem 3 is to shift our attention from lines to axes of symmetry. The number of axes necessary for consideration can be cut down by

THEOREM 4. *If a Bravais net has an axis of n-fold symmetry, it has also n-fold symmetry about any lattice point.*

To prove this, we consider the piercing point A of the axis and an arbitrary lattice point P. By assumption there is a rotation about

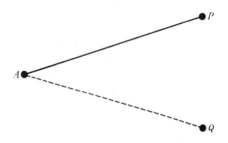

Fig. 1.11. Illustration of Theorem 4 on Bravais nets: axis of symmetry can be shifted to lattice points.

A bringing P to Q which is a symmetry operation (fig. 1.11). Hence, Q is also a lattice point. Therefore, the translation of the net by a vector QP is a symmetry operation by definition of the Bravais net. The result of this operation succeeding the rotation will be a rotation of the lattice about P by an angle PAQ. Hence P is the piercing point of an axis of n-fold symmetry. This reasoning does not exclude, of course, that the symmetry about P may be actually higher than that about A.

We are now ready for the most important theorem of theoretical crystallography:

THEOREM 5. *A Bravais net can have only axes of two-, three-, four- and sixfold symmetry.*

Because of Theorem 4 we can restrict ourselves to axes passing through lattice points, and because of Theorem 1 these axes are necessarily $2n$-fold. Now let O and P be two points in the net

whose distance is the least possible one (fig. 1.12). Then by carrying
out the fundamental rotation about O, we bring P to the lattice
position Q. Now by assumption we must have $\overline{PQ} \geqq \overline{OP}$. This
is only true for $n = 1, 2, 3$. Hence the only possible symmetries
about O are two-, four- or sixfold; and about an arbitrary axis from
Theorem 4, one-, two-, three-, four- or sixfold.

A little more work along the lines
indicated by fig. 1.12 completes the
classification of Bravais nets. The
symmetry about a lattice point O is
necessarily two-, four- or sixfold.
The twofold case has already been
treated. For the other two cases we
draw a circle about O of radius $\overline{OP}$
and place on the circle all images

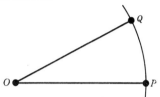

Fig. 1.12. Illustration of Theorem 5
on Bravais nets: an axis of symmetry generates a circular array
of lattice points.

obtained from P by rotation about O (fig. 1.13). We end up with a
square or with a regular hexagon having O as centre; this is shown
in fig. 1.13. No other lattice point could be placed inside the

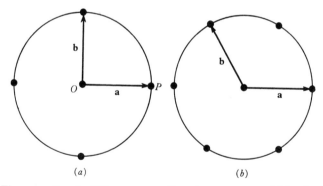

(a)   (b)

Fig. 1.13. Result of Theorem 5 on Bravais nets: a non-trivial axis of
symmetry passing through a lattice point is either four- or sixfold.

circle without violation of the minimum distance premise. Equivalence of all lattice points then permits us to complete the figure.
We end up with the only two high symmetry Bravais nets possible,
the square (fig. 1.14) and the triequiangular (fig. 1.15) nets. The
five possibilities described exhaust therefore all possible Bravais
nets in two dimensions.

Fig. 1.14. The square Bravais net.

Fig. 1.15. The triequiangular Bravais net.

THEORY OF BRAVAIS SPACE LATTICES

If we return now to three-dimensional lattices, we find the collected information still quite relevant. The idea of an n-fold axis certainly needs no elaboration. The line of symmetry has two possible generalizations: either as a twofold axis in the lattice plane or as a symmetry plane at right angles to the plane. Indeed the proper generalization of Theorem 2 is

THEOREM 2 a. *If a Bravais lattice has a twofold axis it also has a plane of symmetry at right angles to that axis and vice versa.*

Theorem 3 is obviously spatial in the first place:

THEOREM 3 a. *If a Bravais lattice has two planes of symmetry making an angle α, the intersection of the two planes is a rotational axis of period 2α.*

Theorems 4 and 5 obviously remain intact except that the jump in the latter from n-fold to $2n$-fold axes by passage to a lattice point is not guaranteed because of the loss of Theorem 1. However, the restriction of Theorem 5 still stands because no lattice can have a symmetry which is impossible in the plane.

The five theorems derived for Bravais nets are a convenient basis for the enumeration of the fourteen Bravais space lattices.

We have already stressed the fact that nothing analogous to Theorem 1 holds in three dimensions. Hence the possibility arises that a Bravais lattice has no planes or axes of symmetry. Such a lattice is called *triclinic*. It is conveniently represented by its primitive cell which is a parallelepiped having neither right angles nor equal sides (fig. 1.16). The next possibility is the presence of a twofold axis. From Theorem 1 it follows then that the plane at right angles need have no further internal symmetry, but according

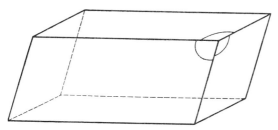

Fig. 1.16. Unit cell of the triclinic Bravais lattice: a general parallelepiped.

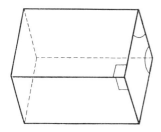

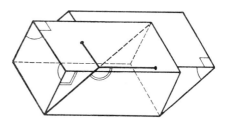

Fig. 1.17. Unit cell of the mono-clinic lattice: a right prism with parallelogram base.

Fig. 1.18. Comparison of the monoclinic unit cell with a primitive cell of the staggered monoclinic lattice: the right angle marked by a double line is replaced by another angle in the primitive cell.

to Theorem 2a the plane is itself a plane of symmetry. If this is the case, we deal with the *monoclinic system*. It may be symbolized by a right prism with parallelogram base (fig. 1.17). However, the cell shown may well have to be larger than the primitive cell because successive layers may be staggered. The phenomenon was discussed already in connection with Theorem 2 and is illustrated in figs. 1.8 and 1.9. Thus we have two Bravais lattices for the monoclinic system, the *staggered* and the *simple* monoclinic lattice.

It is sometimes convenient to maintain the cell of fig. 1.17 even for the staggered form, because the true primitive cell shown in fig. 1.18 obscures the symmetry. In this larger cell, which we call a *unit* rather than a *primitive* cell, extra points appear at other places besides the corners. In the present case there are two points to the unit cell and the extra point may be placed in the body centre by suitable choice of the base.

The next crystal system has two mutually perpendicular planes of symmetry. Theorems 2*a* and 3*a* then yield a third system of planes perpendicular to both and three twofold axes perpendicular to each other. This is the *orthorhombic system* which is representable

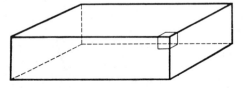

Fig. 1.19. Unit cell of the orthorhombic system:
a rectangular parallelepiped or 'brick'.

by a rectangular parallelepiped (fig. 1.19) or 'brick'. The problem of staggering, with three planes involved, here assumes formidable proportions. We finally find four Bravais lattices in this system, which are respectively a simple brick, a body-centred brick, a brick with all faces centred, a brick with two opposite faces centred (a brick with two pairs of opposite faces centred is not a Bravais lattice). The first two yield rectangular nets in all planes of symmetry, the third a rhombic net in all planes, the last a rhombic net in one, rectangular nets in the two others.

An obvious specialization of the orthorhombic system arises if two axes are made equal. The lattice plane subtended by these two axes is then a square Bravais net (fig. 1.14), and the crystal lattice must have fourfold symmetry (otherwise we would deal with an accidental numerical equality of two nonequivalent lengths, a phenomenon which sometimes occurs in nature in an approximate way). This procedure generates the *tetragonal system*. There are only two ways in which one can stack these square nets and maintain fourfold symmetry. Either one places them all perpendicularly above each other or one staggers them by placing the lattice points of one plane over the centres of the previous squares. In the former case one gets a *simple tetragonal* lattice which has as its primitive cell an upright prism with square base (fig. 1.20). In the latter case one may pick the same prism as a unit cell and place a lattice point in its centre (fig. 1.21); this yields a *body-centred tetragonal* lattice.

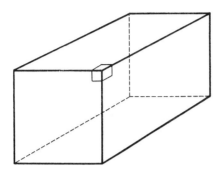

Fig. 1.20. Unit cell of the tetragonal system: a right prism with square base.

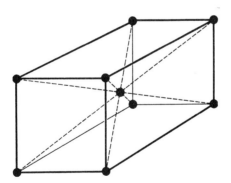

Fig. 1.21. Arrangement in the unit cell of the lattice points of the body-centred tetragonal lattice.

The triequiangular Bravais net shown in fig. 1.15 serves as a base for the *trigonal system* whose characteristic feature is threefold symmetry about an axis. In this system, as in the tetragonal system discussed above, there are two ways of arranging successive layers of the Bravais nets of the type shown in fig. 1.15. Either they are placed perpendicularly above each other, in which case the crystal is said to be *hexagonal*, or else the centre of one triangle is placed perpendicularly above a lattice point of the previous plane; the crystal is then called *rhombohedral*. In the rhombohedral Bravais lattice the original arrangement of lattice points repeats only every

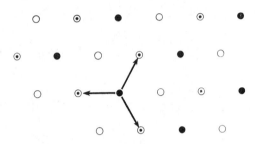

Fig. 1.22. Arrangement of the triequiangular nets in the formation of the rhombohedral lattice; the lattice points of three successive basal planes are marked by full, empty and dotted circles respectively.

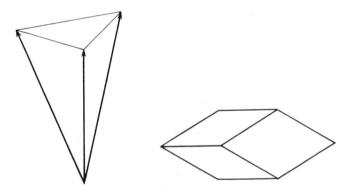

Fig. 1.23. Choice of the three equivalent basis vectors in the rhombohedral lattice.

Fig. 1.24. Unit cell of the rhombohedral lattice: a rhombohedron.

third layer and we have three types of triequiangular planes staggered with respect to each other; fig. 1.22 shows the lattice points of these three planes by full, empty, and dotted circles respectively. The figure shows that the rhombohedral Bravais lattice has lost the sixfold symmetry of its basal plane. This feature is often used as a base for distinguishing a hexagonal and a rhombohedral crystal system; however, many crystals with a hexagonal Bravais lattice also have only threefold symmetry. The choice of the unit cell is conventionalized for the two crystal types as follows. In the rhombohedral lattice an elementary triangle of the basal plane is chosen and the three equivalent vectors from its corners to the lattice point immediately above (or below) it are taken as basis vectors; this is shown in figs. 1.22 and 1.23; the completed cell is then a rhombohedron (fig. 1.24); from this rhombohedron the crystal lattice derives its name. The unit cell so chosen is primitive and exhibits the threefold symmetry of the crystal. No such favourable choice is possible in the hexagonal case because no parallelepiped can exhibit sixfold symmetry. The axes which one chooses conventionally are shown in fig. 1.25.

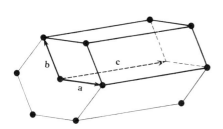

Fig. 1.25. Arrangement of lattice points and conventional choice of basis vectors in the hexagonal lattice; no choice of axes incorporating the hexagonal symmetry is possible.

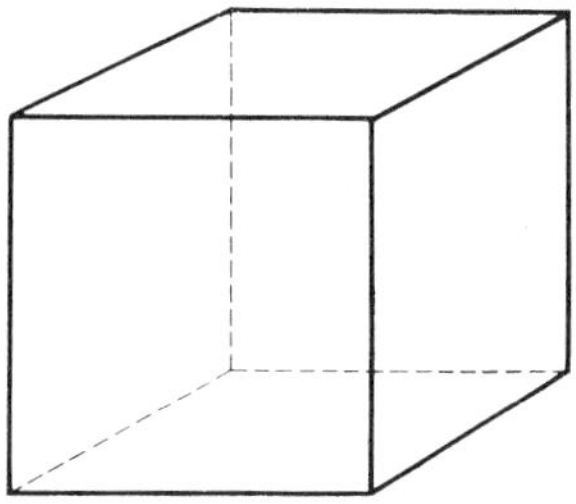

Fig. 1.26. The cube, the unit cell of the cubic system.

In many ways the most interesting crystal system is the *cubic system*, symbolized by the cube as the unit cell, as shown in fig. 1.26. A Bravais lattice which is cubic not only has fourfold symmetry about each side but also threefold symmetry about each body diagonal. Hence it combines features of the tetragonal and the trigonal system. As to the possible Bravais lattices in this system, one realizes there must be at least two, corresponding respectively to the simple (fig. 1.20) and the body-centred (fig. 1.21) tetragonal structures. However, there are actually three cubic Bravais lattices: the *simple*, the *body-centred* and the *face-centred* lattice. Both these latter are specializations of the body-centred tetragonal lattice. This is illustrated in fig. 1.27 where the body-centred tetragonal cell is constructed for the face-centred cubic lattice. Another perspective on these three lattices can be gained by observing that the cubic system is also rhombohedral and that therefore, for all three cubic lattices it must be possible to pick rhombohedral primitive cells. This observation is obvious for the cube which is itself a special kind of rhombohedron. For the face-centred and body-centred lattices the construction of the rhombo-

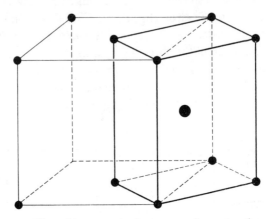

Fig. 1.27. The cubic system is also tetragonal; construction of the body-centred tetragonal cell for the face-centred cubic lattice.

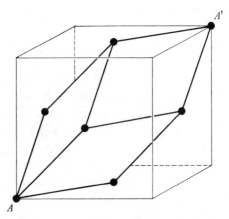

Fig. 1.28. The cubic system is also trigonal; construction of the rhombohedral primitive cell for the face-centred cubic lattice.

hedral primitive cell is carried out in fig. 1.28 and fig. 1.29. $\overline{AA'}$ is the axis of threefold symmetry of the rhombohedron.

A modification of the unit cell concept was introduced by Wigner and Seitz who proposed a cell which is primitive and yet has the symmetry of the lattice.† To construct this cell, one draws connecting lines from one lattice point to all other lattice points and bisects those lines by planes. The Wigner–Seitz cell going with this point is the volume interior to all planes; in other words,

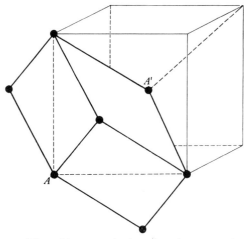

Fig. 1.29. The cubic system is also trigonal; construction of the rhombohedral primitive cell for the body-centred cubic lattice.

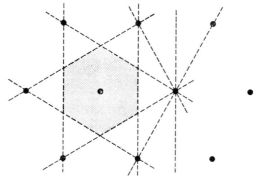

Fig. 1.30. Construction of the Wigner–Seitz primitive cell for the triequiangular Bravais net; the cell is a hexagon enclosing the lattice point.

the Wigner–Seitz cell associated with a lattice point is the locus of all points which are closer to that lattice point than to any other lattice point. Fig. 1.30 illustrates this in the plane for the tri-equiangular net, where the Wigner–Seitz cells are small hexagons

† E. Wigner and F. Seitz, *Phys. Rev.* **43** (1933), 804.

with the lattice points as centres. Fig. 1.31 shows the Wigner–Seitz cell for the body-centred cubic lattice. It is a truncated octahedron; the octahedral planes are determined by nearest neighbours to the central point which are at the eight cube corners shown, and the truncation is produced by the next nearest neighbours in the centres of the six adjoining cubes.

The Bravais lattices just discussed, beside allowing a simple classification of crystal types, are also useful in labelling lattice

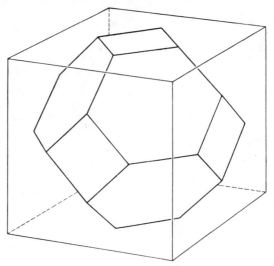

Fig. 1.31. Wigner–Seitz cell for the body-centred cubic lattice:
a truncated octahedron.

vectors, lattice directions, and lattice planes. For this purpose the choice of the three basis vectors **a**, **b**, **c** of a structure is first fixed by convention; thereupon, the labelling of positions is carried out with the help of equation (1.2). Integer values of the three numbers l, m, n (and also simple fractions such as $(\frac{1}{2}, \frac{1}{2}, 0)$ if the unit cell is larger than a primitive cell) are equivalent to $(0, 0, 0)$. The interior of the cell may be explored by giving l, m, n values between 0 and 1. For planes the most obvious procedure would be to give the intercepts on the three axes in terms of the basis vectors **a**, **b**, **c** generating the lattice. Interest is in this case primarily in lattice planes, which appear as external faces or reflecting planes for X-rays. Three integers can be associated with these planes in the following way. Suppose we are originally given the plane by three lattice vectors **u**, **v**, **w** which reach it from the origin. Let the vectors have components as follows

$$\mathbf{u} = (l, \ m, \ n),$$
$$\mathbf{v} = (l', \ m', \ n'),$$
$$\mathbf{w} = (l'', m'', n'').$$

The numbers $l, m, \ldots$ are either integers or very simple fractions. A general vector reaching the plane is then given by

$$\mathbf{x} = \mathbf{u} + \kappa(\mathbf{v} - \mathbf{u}) + \lambda(\mathbf{w} - \mathbf{u}),$$

where κ and λ are arbitrary, or

$$\mathbf{x} = \alpha\mathbf{u} + \beta\mathbf{v} + \gamma\mathbf{w}$$

with $$1 = \alpha + \beta + \gamma.$$

Writing this out in terms of components we get

$$x = \alpha l + \beta l' + \gamma l'',$$
$$y = \alpha m + \beta m' + \gamma m'',$$
$$z = \alpha n + \beta n' + \gamma n'',$$
$$1 = \alpha + \beta + \gamma.$$

Elimination of α, β, γ yields

$$\begin{vmatrix} x & l & l' & l'' \\ y & m & m' & m'' \\ z & n & n' & n'' \\ 1 & 1 & 1 & 1 \end{vmatrix} = 0. \tag{1.5}$$

This is an equation of the form

$$Lx + My + Nz = D, \tag{1.6}$$

in which all coefficients are or can be made integers. The triple (LMN) characterizes the inclination of the plane; if it is the smallest set of integers possible, the three integers are called the *Miller indices* of the plane; they are inversely proportional to the intercepts along the three axes, as is evident from (1.6). Fig. 1.32

shows the orientation of the three most important planes in the cubic system. In this case the Miller indices coincide with the notion of direction parameters used in solid geometry. The converse of the statement just proved is also true with certain restrictions †.

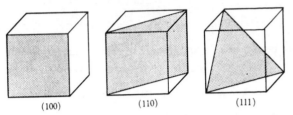

$$(100) \qquad\qquad (110) \qquad\qquad (111)$$

Fig. 1.32. Unit cubes exhibiting the three simple lattice planes of the cubic system: the (100), (110) and (111) plane.

It is important to realize that low index lattice planes have generally a greater density of lattice points than high index ones. We prove this for the simple cubic lattice by treating L, M, N in (1.6) as fixed and D as variable. We have then a set of parallel equidistant planes, for which $D = 0, 1, 2, 3, 4, \ldots.$ By the rules of solid geometry, the distance of successive planes is:

$$d = \frac{a}{(L^2 + M^2 + N^2)^{\frac{1}{2}}}, \qquad (1.7)$$

where a is the length of the fundamental translation along the cube edge. This formula shows that high index planes are stacked more closely than low index ones. Now, the density of lattice points on the planes must be proportional to the mutual distance of these planes, because a unit area of plane will always cover the same number of lattice points in the mean when travelling unit distance; hence, if positions containing lattice points are reached more often, the number of lattice points found each time must be correspondingly smaller. Calibration with the (100) plane shows that the density of lattice points in the simple cubic lattice is:

$$\rho = \frac{1}{a^2 (L^2 + M^2 + N^2)^{\frac{1}{2}}}. \qquad (1.8)$$

† See problem 4 at the end of the chapter.

Modifications occur for the other cubic lattices. For the face-centred cubic structure, for instance, the (111) plane is the plane of densest packing; this is exhibited in fig. 1.35 below.

EXAMPLES OF CRYSTAL STRUCTURES

If we wish to advance from discussing the symmetry of Bravais lattices to the discussion of real crystals we must imagine each primitive cell filled with identical material, and thereupon proceed afresh with an examination of the resultant structure. Only the translational symmetry of this new structure is necessarily identical with that of the underlying Bravais lattice. Other elements of symmetry deduced for the latter may or may not be maintained in the complete structure. Thus the fourteen space lattices just described become diversified into 230 complete space groups. Each of these groups can be the symmetry type of several different crystal structures, and each crystal structure can be realized by several different chemical compounds (substances exhibiting this type of similarity are called *isomorphous*, if the angles and distances are also very similar; mixed crystals are then apt to occur). The study of all this wealth is clearly outside the scope of this book. We shall be satisfied here to pass in review some important examples and to add a few observations of principle.

The most obvious way of creating a crystal structure from a Bravais lattice consists in replacing the lattice points by atoms having themselves (potentially) a higher symmetry. The result will be a structure equivalent in symmetry to the Bravais lattice. Very few crystals are of this type. The two most common examples are the *body-centred* and *face-centred cubic structures*. They are exhibited in figs. 1.33 and 1.34 in such a way that one atom is placed in the centre of the cubic unit cell. All its nearest neighbours are then located on the surface of the cube as shown; the number of these neighbours is read off as being 8 and 12 respectively for the two structures. These numbers are in excess of normal valencies. Therefore, one finds the two structures realized for solids whose cohesive forces have little or no valence character (see Chapter 8), notably metals and the solid phases of the noble gases. The nature of these structures can be made evident by thinking of the atoms as sticky hard spheres. From this point of view both the arrangements leave very few voids. In fact, the face-centred cubic structure is also called *cubic close packed* because it is one of the two densest packings which can be achieved for hard spheres. This aspect of the face-centred cubic structure is brought out in fig. 1.35, where a cubic block of such spheres is first shown and then the same picture with a cutaway view of the close-packed (111) plane.

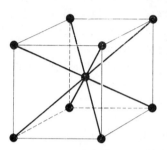 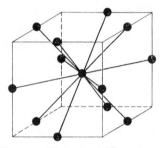

Fig. 1.33. Nearest-neighbour arrangement in the body-centred cubic structure: one atom is surrounded equally by eight others arranged in a cube.

Fig. 1.34. Nearest-neighbour arrangement in the face-centred cubic structure: one atom is surrounded equally by twelve others arranged on the edges of a cube.

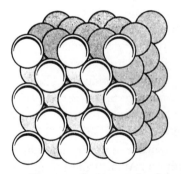 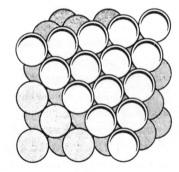

Fig. 1.35. Packing study of the face-centred cubic structure: elementary cube and cube with cut away (111) plane.

A second possibility of proceeding from lattices to crystals consists in introducing several types of atoms to fill the cell, but placing each atom at a position of full symmetry. A case of this type is the *rocksalt structure* (fig. 1.36). Its Bravais lattice is face-centred cubic. Both the sodiums and chlorines separately form such a lattice, shifted by half a cube edge. Each atom is on a site of full cubic symmetry, hence the original symmetry of the lattice is preserved. A similar situation prevails in the *caesium chloride structure*, where the Bravais lattice is simple cubic and the two ions are shifted with respect to each other by half a cube diagonal (fig. 1.37). The structures shown here resemble the two preceding ones

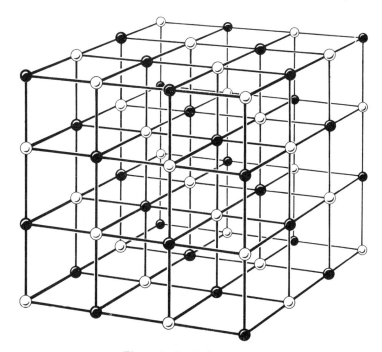

Fig. 1.36. Rocksalt structure.

in that each atom is surrounded by a large number of nearest neighbours (6 and 8 respectively); these neighbours are now of the opposite kind. The structures are therefore realized for ionic *AB* compounds in which the ions *A* and *B* bear opposite charges and thus attract each other, while equal ions repel, and where the forces between the two constituents do not saturate.

We cover already a wider class of structures if we fill the Bravais lattice in such a way that some atoms occupy positions of symmetry lower than that of the lattice, but have them go over into each other under the symmetry operations of the lattice. The *perovskite* and *arsenic structures* may serve as examples for this. Perovskite, $SrTiO_3$, is cubic with the atoms as shown in fig. 1.37, where strontium occupies the corners, titanium the body centre and the oxygens (not shown) the centres of all faces. Clearly these last ones are in a tetragonal rather than cubic environment, but the crystal as a whole yet preserves the symmetries of a cube. Similarly, arsenic and its homologues, antimony and bismuth, have a rhombohedral structure with two atoms per unit cell. In fig. 1.38, the atoms are placed at equal distances above and below the Bravais lattice points, on the threefold axis; the structure retains then the reflection symmetry of the lattice at any lattice point. This symmetry is the

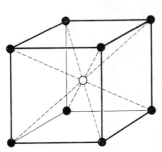

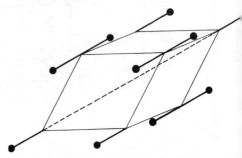

Fig. 1.37. Unit cell of the caesium chloride structure.

Fig. 1.38. Rhombohedral unit cell of arsenic with paired atomic arrangement.

only one missing at the sites of the bismuth atoms themselves. To account for this particular arrangement of atoms, emphasis must be shifted to the linkage of the atoms rather than the symmetry of the crystal. This is done in the illustration of fig. 1.39. The first intuitive impression of the structure is that of a simple cube. Each atom has therefore six neighbours. However, the cube is distorted; the distortion is such that the pseudocubic bonds are divided into long and short ones. The short ones are shown in darker shading in fig. 1.39; they connect each atom to three neighbours, in accordance with the valence of these semimetal atoms. If only these bonds are counted, the crystal consists of layers of covalently bonded atoms spreading at right angles to the trigonal axis.

The normal relationship between the Bravais lattice and the true crystal structure is that some of the symmetries of the lattice are not present in the crystal or are only retained in an attenuated form. As the first example let us choose the second crystalline structure into which sticky hard spheres can be packed, the *hexagonal close packed structure*. It is the third of the common crystalline forms for metals. The arrangement of atoms in the basal plane is the one shown in fig. 1.15. It is evident from fig. 1.35 that this is also

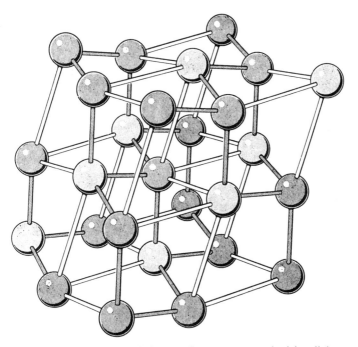

Fig. 1.39. Alternative view of the arsenic structure emphasizing linkage: a slightly distorted simple cubic structure; the shortest link is shaded; the slightly lighter balls outline the rhombohedral cell.

the arrangement of atoms in the (111) plane of the cubic close-packed structure. The difference must arise therefore in the stacking of these layers. The stacking for the cubic case was shown in fig. 1.22; it destroys the sixfold symmetry of fig. 1.15, but retains thereby many other symmetry elements. In the close packed hexagonal structure the difference arises in the placing of the third layer, which simply repeats the first as shown in fig. 1.40. The displacement from a full circle to the closest open circle is now not a crystal period, and the vector from one full circle to another directly above it becomes the smallest lattice vector outside the basal plane. We have therefore a hexagonal Bravais lattice as shown in fig. 1.25. However, the sixfold symmetry exists only in a reduced form, as is shown by the dotted hexagon in fig. 1.40. The open circles can be made to coincide with the full circles by a rotation of 60° about the centre (marked by a star) and *simultaneous translation along the hexagonal axis* by half a period: in other words, we have a *hexagonal screw axis* which still remains a normal threefold axis.

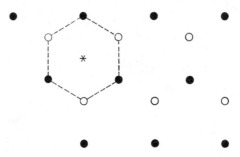

Fig. 1.40. Hexagonal close-packed structure; two layers, shown as empty and full circles, alternate; the dotted hexagon and the star indicate a sixfold screw axis.

Another modification of hexagonal symmetry arises in the *tellurium structure*, which is also one of the possible forms of selenium. The structure is illustrated in fig. 1.41. Each layer of atoms in the basal plane forms a simple triequiangular net of the type shown in fig. 1.15. The layers are repeated with an offset such that the original arrangement is repeated only every third layer. In fig. 1.41 these three layers are shown as full, open, and dotted circles respectively. From the point of view of symmetry we have here a structure which has no sixfold symmetry in any sense, since opposite directions along one of the dashed lines are non-equivalent; yet the Bravais lattice is hexagonal. The threefold symmetry shows up as threefold screw axes located at the intersections of the lines or the centres of the triangles. What distinguishes these screw axes from the previous axes is that they have here a definite screw sense; this gives rise to two non-equivalent forms which are mirror images of each other. From the point of view of bonding, the tellurium structure represents the opposite extreme from the close packed hexagonal structure. Each atom has only two nearest neighbours in accordance with the valence of selenium and telurium. The bonded units form twisted strands

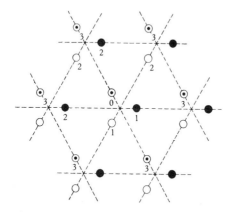

Fig. 1.41. Tellurium structure. Three layers, shown as open, full and dotted circles, alternate.

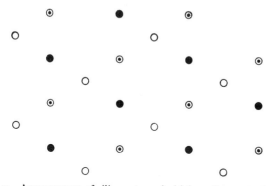

Fig. 1.42. Arrangement of silicon atoms in high or β quartz; the three layers are shown as full, open and dotted circles.

winding around the threefold axes. The strands are well separated and the bonding between them is almost certainly much weaker than the valence bonding within them.

Still another modification of hexagonal symmetry arises in the *β quartz structure*, whose three layers of silicon atoms are shown in fig. 1.42. (The oxygens may be taken for this discussion simply as half way between any two nearest neighbours; actually, they are pushed out of this alignment into the voids beside them). The crystal has sixfold screw axes located in the centres of the empty hexagons; they are true twofold axes, but their threefold character has become screw type. The screw axes have again a definite sense. Hence quartz will have two distinct mirror formations and any one single crystal will distinguish between right-hand

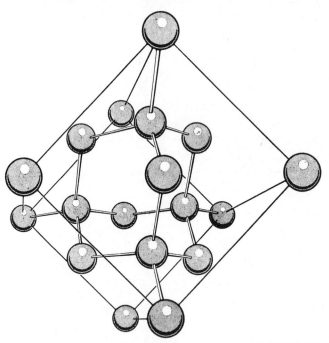

Fig. 1.43. Cubic unit cell of the diamond structure (drawn by F. M. Thayer).

and left-hand polarized light travelling along its axis. As a structure quartz can serve as an example of a crystal bonded throughout by chemical valence. Each oxygen is placed at equal distance from the two nearest-neighbour silicon atoms; it has therefore only two silicon neighbours. Each silicon has four oxygen neighbours. Thus a single valence bond can be drawn exactly from neighbour to neighbour. This is the feature typical of the chemical species. There are two other crystalline forms of SiO_2, tridymite and cristobalite, whose crystalline symmetries are quite different from quartz.

Yet the nearest-neighbour relationships are the same. In fact, even amorphous silica glass has essentially the same nearest-neighbour arrangement. Quite generally, close neighbour relationships are more typical of a chemical species than overall symmetries. Another interesting feature concerning the SiO_2 structures are the large voids appearing in the crystal and which show up in fig. 1.42 as empty hexagons. Their presence makes quartz the very opposite of a close packed crystal. This is the general rule for valence-bonded crystal structures.

As examples of reduced symmetry in the cubic system, we take the *diamond* and *zincblende structures*. The lattice is shown in fig. 1.43. It is seen to consist of two face-centred cubic lattices which are shifted with respect to each other by a quarter of a cube diagonal. In diamond (and also silicon and germanium) all atoms are of the same kind, while in zincblende one lattice consists of zinc, the other of sulphur. The Bravais lattice is therefore face-centred cubic. One observes that the threefold axes are preserved in this type of structure, but that the fourfold symmetry has disappeared. To observe the screw symmetry which has replaced it, we project the contents of fig. 1.43 on the front face, representing the four layers respectively by open circles, full circles, open squares, full squares (fig. 1.44). The screw axes lie in the centre of the squares, one of which is outlined. No screw sense is to be attributed to the crystal as a whole because screws of opposite type adjoin directly. This reasoning preserves the fourfold axis as a screw axis for diamond, but not for zincblende because open and full points are of different kinds. The axis remains as a fourfold axis of rotation-reflection which one gets by turning fig. 1.44 about its centre by 90° and then reflecting everything on the front face; this last

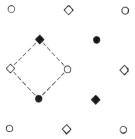

Fig. 1.44. Projection of the contents of the unit cell of diamond on a cube face; the four layers are shown as open circles, full circles, open squares and full squares respectively.

operation interchanges full circles and squares but leaves the open points unaltered; this complicated procedure restores the original arrangement. The zincblende, tellurium and quartz structures are, by the way, the only lattices listed here which have no centre of symmetry and for which, therefore, opposite directions of the same axis are not equivalent. This is particularly easily seen in fig. 1.45 in which one of the (111) directions of zincblende is vertical and the cubic cell is suppressed. One sees that all vertical connecting lines have a full circle at the top and an open circle at the bottom. The (111) and ($\overline{1}\overline{1}\overline{1}$) directions are therefore not equivalent. This property of crystals can sometimes be discerned by macroscopic methods; it

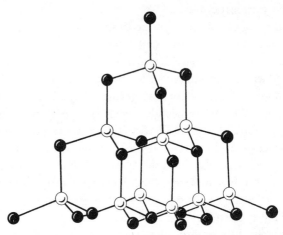

Fig. 1.45. Picture of the diamond or zincblende structure emphasizing linkage; each atom has four nearest neighbours; in zincblende the open and full atoms are different.

shows up by unequal development or appearance of opposite faces. The phenomenon is called *hemihedrism*. Thus, for instance, the tetrahedron can be taken as an extreme hemihedric development of the octahedron (fig. 1.46). Zincblende might therefore be expected to show this type of hemihedric character. From the point of view of nearest-neighbour arrangement, the diamond-zincblende structure is a valence structure in which each atom is surrounded tetrahedrally by four neighbours. The arrangement of these tetrahedra within the cubic unit cell is shown in fig. 1.43, and the resultant tetrahedral linkage is shown in fig. 1.45. This type of structure is obviously fitted to carbon, silicon, and germanium because it permits the placing of just one valence bond between any two neighbours. Its appearance for ZnS is, however, somewhat surprising; no attempt will be made here to explain the structure. The stubbornness of this particular feature of zinc sulphide (even though unexplained) will be appreciated by the reader if he is told that it occurs in a second crystalline modification of ZnS, wurtzite, which stands in the same relationship to the hexagonal close packed lattice as zincblende to the cubic close packed lattice. A different symmetry is produced, but again, each zinc atom is surrounded tetrahedrally by four sulphurs, and each sulphur by four zincs.

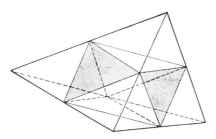

Fig. 1.46. Development of hemihedric forms; the tetrahedron as an extreme hemihedric development of the octahedron.

It is hoped that these few examples will be sufficient for a first rough study of solid state theory. The student will find it advantageous to link deeper penetration of the field with a more ample knowledge of geometric relationships.

PROBLEMS

1. Make a drawing demonstrating how the vector (2, 3) of a plane Bravais net can be used as one of its basis vectors. What second basis vector have you chosen? Show that the area of the parallelogram subtended by the vector (2, 3) and your chosen vector is equal to the area of the one subtended by the vectors (1, 0) and (0, 1).

2. Prove that after choosing the first basis vector **a** of a Bravais lattice arbitrarily you can pick the second one so that it subtends together with **a** any pre-assigned lattice plane containing **a**.

3. Construct a system of basis vectors of a Bravais lattice, using as one of them the vector (2, 3, 4).

4. Prove that any equation of the form (1.6) defines a lattice plane if L, M, N and D are integers such that D is divisible by the greatest common divisor of the other three.

5. Since the cubic unit cell of the body-centred cubic lattice contains two atoms per cell, it follows that the volume of any primitive cell must equal one half the cube volume. Prove this explicitly for the Wigner-Seitz cell shown in fig. 1.31.

6. Find the Wigner–Seitz cell of the face-centred cubic lattice.

7. Suppose hard sticky spheres are packed into a crystalline arrangement such as the face-centred cubic structure of fig. 1.35. Calculate the fraction of the crystal volume occupied by the spheres for this case, and also for the body-centred, simple cubic and diamond structures. Comment on the numbers obtained.

Fig. 1.47. A plane pattern having an unusual symmetry.

8. Find the c/a ratio of a close-packed hexagonal lattice of hard spheres and compare it with empirical ratios of materials found in nature. Find also the fraction of the crystal volume occupied by these spheres. Compare the answer with the results of problem 7.

9. Prove that the body-centred and face-centred rhombohedral lattices are simple rhombohedral.

10. Since there is no metric distinction between a rhombohedral and a hexagonal lattice, hexagonal unit cells can be found for a rhombohedral lattice and rhombohedral cells for a hexagonal lattice. Construct the smallest such cell in each case and determine the ratio of its volume to that of the primitive cell.

11. Find the lattice plane with the densest packing in the body-centred cubic structure.

12. Theorists often prefer to study hexagonal crystals with the help of three mutually perpendicular axes, which means that they treat them as orthorhombic. Of which of the four orthorhombic space lattices is the hexagonal Bravais lattice a special case, and what special axis ratio does it entail?.

13. Plot the curves $y = \sin x$ and $y = |\sin x|$ in rectangular cartesian co-ordinates and consider them as symmetric graphs. Find for each the symmetry operations which leave it invariant, excluding operations involving the third dimension. Reduce the symmetry group to a generating basis, that is, pick the smallest possible number of elements which by repetition, combination or inversion, generate all others.

14. Carry out the same analysis as in problem 13 for the pattern in the adjoining fig. 1.47, when viewed as part of an infinite plane graph.

15. View the patterns of fig. 1.40 and 1.41 as plane infinite graphs and ignore the distinction of open, full and dotted circles. Derive for each graph its symmetry group; reduce the group to its generating elements.

16. Derive the space group of the perovskite, diamond and zincblende structures; reduce each group to a generating basis as explained in problem 13.

17. The text indicates that wurtzite is a four-coordinated structure which stands in the same relation to the close packed hexagonal structure as diamond to the close packed cubic structure. Make use of this hint to give the position of all atoms in the unit cell shown in fig. 1.25.

18. Analyse the symmetry group of the cube. Show that the number of elements in the group is 48, and that it has 10 classes. (A class comprises all elements which are geometrically congruent, but distinguished through the location of the axis of rotation, plane of symmetry, etc. If, and only if, A and B are members of the same class, a relationship $A = CBC^{-1}$ exists, with C any element of the group.)

BIBLIOGRAPHY

ELEMENTARY AND DESCRIPTIVE WORKS

C. W. BUNN. *Chemical Crystallography.* (Oxford, Clarendon Press, 1945.)
K. LONSDALE. *Crystals and X-rays.* (London, Bell and Sons, 1948.)
L. PAULING. *Nature of the Chemical Bond and the Structure of Molecules and Crystals,* 2nd ed. (Ithaca, Cornell Univ. Press, 1945.)
F. C. PHILLIPS. *Introduction to Crystallography,* 2nd ed. (New York, Longmans, Green, 1956.)

THEORETICAL TREATISES

H. HILTON. *Mathematical Crystallography and the Theory of Groups of Movements.* (Oxford, Clarendon Press, 1903.)
A. SCHOENFLIES. *Theorie der Kristallstruktur.* (Berlin, Borntraeger, 1923.)
W. VOIGT. *Lehrbuch der Kristallphysik.* (Leipzig, Teubner, 1910.)

F. Seitz. A matrix-algebraic development of crystallographic groups. *Z. Kristallogr.* **88** (1934) 433–59; **90** (1935), 289–313; **91** (1935), 336–66; **94** (1936), 100–30.

REFERENCE WORKS

J. D. H. Donnay and W. Nowacki. *Crystal Data.* (New York, Geol. Soc. of America, 1954.)

International Union of Crystallography. *International Tables for X-ray Crystallography.* (Birmingham, England, Kynoch Press, 1952. Three vols.)

C. Palache, H. Berman and C. Frondel. *System of Mineralogy of James Dwight Dana and Edward Salisbury Dana.* Yale University, 7th ed. (New York, Wiley 1951. Two vols.)

A. N. Winchell. *Microscopic Characters of Artificial Inorganic Solid Substances or Artificial Minerals*, 2nd ed. (New York, Wiley, 1931.)

R. W. G. Wyckoff. *Crystal Structures.* (New York, Interscience, 1948–53. Three vols.)

CHAPTER 2

DIFFRACTION OF RADIATION BY CRYSTALS

Physics is fortunate in having an almost perfect tool to verify the geometrical notions of Chapter 1, namely the diffraction of X-rays by crystals; this tool can be supplemented by diffraction of electrons and neutrons.

The essential condition for exploring the atomic arrangement of crystals by interference methods is to have at one's disposal waves whose wave length is comparable to the interatomic distances. X-rays are electromagnetic waves of this type. For material particles the energy E must be picked according to De Broglie's relation

$$E = \frac{h^2}{2\,m\,\lambda^2},$$
(2.1)

where h is Planck's constant, m the particle mass and λ the wave length. This yields for electrons an energy of a few electron volts, and for neutrons thermal energy. The usefulness of most waves for this purpose is limited by their small depth of penetration; for the coherent scattering phenomenon of the ordered solid can be observed only if the identical scattering units are large in number, form a three-dimensional array, and all contribute about equally to the scattered signal. These requirements exclude all but the three wave motions mentioned above; even among them, electron diffraction is useful only in special cases because of the limited penetrating power of low-energy electrons.

Wave theory permits us to write down a solution for the following generalized problem. Let there be a number of identical scattering centres displaced with respect to each other in space; let their position vectors be $\mathbf{r}$. Let there be an incoming wave of wave vector $\mathbf{k_0}$ and amplitude A, and let each centre produce an equal scattered wave coherent to the incoming signal; let its amplitude be fA where f depends on the wave length $1/k_0$ and the scattering angles, but not on A. Then, if we look in a direction specified by a wave vector $\mathbf{k}$ for the total scattered amplitude A' from all centres, it equals

$$A' = Af \sum_{\mathbf{r}} e^{2\pi i (\mathbf{k} - \mathbf{k_0}) \cdot \mathbf{r}}.$$
(2.2)

For the case of diffraction by crystals the equivalent scattering units are the primitive cells (or occasionally approximately equivalent sub-portions of the primitive cell). f is then called the *form factor*; it will not be discussed any further here. The sum in (2.2)

$$S = \sum_{\mathbf{r}} e^{2\pi i(\mathbf{k}-\mathbf{k}_0)\cdot\mathbf{r}} \qquad (2.3)$$

is called the *structure factor*; it is generally a sum over the points of a Bravais lattice, that is $\mathbf{r}$ takes on all values permitted by (1.2) within some range.

In order to understand the nature of the structure factor S, let us carry out the sum (2.3) explicitly for the case of a crystal whose bounding planes are parallel to the bounding planes of the chosen unit cell. The indices l, m, n in (1.2) run then from 1 to fixed upper limits L, M, N. The sum in (2.3) is then a product of three geometric series and yields, apart from an unimportant phase constant,

$$S = \frac{\sin \pi L(\mathbf{k}-\mathbf{k}_0)\cdot\mathbf{a}}{\sin \pi(\mathbf{k}-\mathbf{k}_0)\cdot\mathbf{a}} \times \frac{\sin \pi M(\mathbf{k}-\mathbf{k}_0)\cdot\mathbf{b}}{\sin \pi(\mathbf{k}-\mathbf{k}_0)\cdot\mathbf{b}} \times \frac{\sin \pi N(\mathbf{k}-\mathbf{k}_0)\cdot\mathbf{c}}{\sin \pi(\mathbf{k}-\mathbf{k}_0)\cdot\mathbf{c}}. \qquad (2.4)$$

It is seen that the structure factor is the product of three factors of the type

$$S^* = \frac{\sin \pi L x}{\sin \pi x}, \qquad (2.5)$$

where x is a fixed function of the wave lengths, angles and cell dimensions, and L is the number of coherently scattering ordered units along some direction. If L is large, S^* is a rapidly oscillating function of x with many zeros and of order of magnitude not more than 1. Exceptions to this rule arise at the zeros of the denominator where S^* rises to its maximum value L. The width in x of these peaks is about $2/L$. The expression S^* is, in fact, the well-known grating amplitude function for a grating of L lines, and the maxima arise when the scattered wavelets from each grating slit are in phase. Since the crystal is a three-dimensional grating, the function S is the product of three such factors. If the condition is satisfied that each of the factors is at its maximum then S is above its background value by a factor which equals the number of cells in the crystal. This could have been inferred directly from (2.3) because all terms in the sum S are then in phase. The conditions for such a maximum are

$$\left.\begin{array}{l} (\mathbf{k} - \mathbf{k_0}) \cdot \mathbf{a} = L, \\ (\mathbf{k} - \mathbf{k_0}) \cdot \mathbf{b} = M, \\ (\mathbf{k} - \mathbf{k_0}) \cdot \mathbf{c} = N, \end{array}\right\} \qquad (2.6)$$

where L, M, N are integers. These equations are the so-called *Laue conditions* for X-ray diffraction.

In interpreting the equations (2.6) one has to comprehend the qualitative difference of the interference phenomenon studied here as compared to one from a grating of lower dimension. If the grating had only one or two dimensions then the equations (2.6) could, for fixed wave length and fixed orders L and M, be solved for the two angles of scattering contained in the left-hand side. In three dimensions, however, we deal with three equations in two unknowns, which have generally no solution. There are two ways out of this difficulty. One, the so-called Laue method, uses radiation covering a wide band of wave lengths; the other, the so-called Bragg method, rotates the crystal mechanically about some axis, but uses X-rays of fixed wave length. In the first method the Laue equations determine the two angles of scattering and the wave length, in the second the two scattering angles and the positional azimuth of the crystal. For either procedure, a thorough understanding of the equations (2.6) is needed, and this is best obtained from the Bragg interpretation of the equations.

In Bragg's interpretation of the equations (2.6) two out of the three equations are summed up by the observation that the direction of the scattered light is as if it had been reflected by the lattice plane of Miller indices $(L/\sigma, M/\sigma, N/\sigma)$ where σ is the greatest common divisor of L, M, N. There remains, then, one equation which is a condition on the wave length or the angle of incidence on this plane.

To prove the first part, we consider two points of co-ordinates (x_1, y_1, z_1) and co-ordinates (x_2, y_2, z_2) lying on the same lattice plane $(L/\sigma, M/\sigma, N/\sigma)$. Each point obeys then the same equation (1.6), namely

$$Lx + My + Nz = D,$$

and hence

$$L(x_2 - x_1) + M(y_2 - y_1) + N(z_2 - z_1) = 0.$$

This equation, together with (2.6), means that the vector $\mathbf{k} - \mathbf{k}_0$ is perpendicular to the plane. For

$$\begin{aligned}
(\mathbf{k} - \mathbf{k}_0) \cdot (\mathbf{r}_2 - \mathbf{r}_1) &= (\mathbf{k} - \mathbf{k}_0) \cdot [(x_2 - x_1)\,\mathbf{a} + (y_2 - y_1)\,\mathbf{b} + (z_2 - z_1)\,\mathbf{c}] \\
&= L(x_2 - x_1) + M(y_2 - y_1) + N(z_2 - z_1) \\
&= 0.
\end{aligned}$$

Since the vectors are also equal in length they make equal angles with their difference. Hence they make equal angles with the normal to the lattice plane and lie in the same plane with it; this is the law of reflection (fig. 2.1).

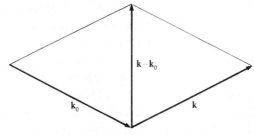

Fig. 2.1. Geometry of wave vectors in Laue–Bragg scattering.

The remaining information in (2.6) is best extracted by calculating the phase difference between reflections from successive lattice planes. Let $\mathbf{r}_1$ be located on one plane and $\mathbf{r}_2$ on the next one. Then the co-ordinates obey the equations

$$Lx_1 + My_1 + Nz_1 = D,$$
$$Lx_2 + My_2 + Nz_2 = D + \sigma,$$

where σ is the greatest common divisor of L, M, N. Hence

$$L(x_2 - x_1) + M(y_2 - y_1) + N(z_2 - z_1) = \sigma.$$

Now the phase going with the scattering from a point with position vector $\mathbf{r}$ is read off from (2.2) as $2\pi(\mathbf{k} - \mathbf{k}_0) \cdot \mathbf{r}$. Hence the phase difference between points of position vectors $\mathbf{r}_2$ and $\mathbf{r}_1$ is

$$2\pi(\mathbf{k} - \mathbf{k}_0) \cdot (\mathbf{r}_2 - \mathbf{r}_1).$$

This, together with (2.6) yields

$$\begin{aligned}
\text{Phase difference} &= 2\pi\,(\mathbf{k} - \mathbf{k}_0) \cdot (\mathbf{r}_2 - \mathbf{r}_1) \\
&= 2\pi\,(\mathbf{k} - \mathbf{k}_0) \cdot [(x_2 - x_1)\,\mathbf{a} + (y_2 - y_1)\,\mathbf{b} + (z_2 - z_1)\,\mathbf{c}] \\
&= 2\pi\sigma;
\end{aligned}$$

that is, the phase difference is a σ-fold multiple of 2π. In the language of optics, σ is the 'order' of the spectrum. The statement about phase difference can be given a geometrical content with the help of the drawing of fig. 2.2. The drawing shows the incident and reflected rays from two successive lattice planes at a distance d. The path difference between them must be σ times the wave length. Hence we read off immediately from the figure

$$2d\cos i = \sigma \lambda, \qquad (2.7)$$

where i is the angle of incidence of the radiation. This is *Bragg's Law*. It, together with the law of specular reflection, is equivalent to the interference conditions (2.6). As anticipated, the law yields the

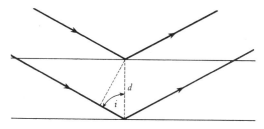

Fig. 2.2. Laue–Bragg scattering viewed as reflection from planes: derivation of Bragg's law.

angle of incidence as function of the wave length, or vice versa, for each reflecting plane and each order. It is seen that the structure of the equation is such that high-order lattice planes (which have a small d) or high-order reflections (high σ) for low-order planes are automatically eliminated if λ is of the order of inter-atomic dimensions.

The reasoning which leads from (2.3) to (2.6) stands as given if the indices L, M, N refer to a primitive cell. However, X-rays are used to determine crystal structure; for this and other reasons reference will often be made to a cell deduced from the crystal shape; such a cell will have the crystal symmetry and is apt to be a unit rather than a primitive cell. In consequence, if the Bravais lattice is body-centred with respect to the cell used, two terms per cell will appear in the structure factor, which differ from each other by a factor $e^{2\pi i(\mathbf{k}-\mathbf{k_0})\cdot[\frac{1}{2}\mathbf{a}+\frac{1}{2}\mathbf{b}+\frac{1}{2}\mathbf{c}]}$.

The remainder of the sum proceeds as previously, yielding the conditions (2.6). With the help of these conditions the above factor can be simplified, yielding finally

$$S_{\text{body centred}} = S_{\text{simple}}[1 + e^{\pi i(L+M+N)}]. \qquad (2.8)$$

The extra factor makes S vanish whenever $L+M+N$ is odd; this means that all odd-order reflections vanish from planes for which the sum of the Miller indices is odd; the reason is clearly that the reflections from the corners and the centres of the unit cells are out of phase for these reflections. Such missing reflections can be used to determine not only the exact nature of the Bravais lattice, but also to detect the presence of screw axes, glide planes, etc.

Manipulation of (2.6) is of such importance that it seems worthwhile in this instance to give a third form of interpretation of them in terms of the reciprocal lattice. Suppose there is given a space in terms of its three generating basis vectors $\mathbf{a}, \mathbf{b}, \mathbf{c}$. Then its *reciprocal space* is generated by three basis vectors $\mathbf{a}^*, \mathbf{b}^*, \mathbf{c}^*$ which obey the prescriptions

$$\left.\begin{array}{lll} \mathbf{a}^* \cdot \mathbf{a} = 1, & \mathbf{a}^* \cdot \mathbf{b} = 0, & \mathbf{a}^* \cdot \mathbf{c} = 0, \\[4pt] \mathbf{b}^* \cdot \mathbf{a} = 0, & \mathbf{b}^* \cdot \mathbf{b} = 1, & \mathbf{b}^* \cdot \mathbf{c} = 0, \\[4pt] \mathbf{c}^* \cdot \mathbf{a} = 0, & \mathbf{c}^* \cdot \mathbf{b} = 0, & \mathbf{c}^* \cdot \mathbf{c} = 1. \end{array}\right\} \qquad (2.9)$$

The equations (2.9) are solved explicitly through

$$\mathbf{a}^* = \frac{\mathbf{b} \times \mathbf{c}}{\mathbf{a} \cdot (\mathbf{b} \times \mathbf{c})}; \quad \mathbf{b}^* = \frac{\mathbf{c} \times \mathbf{a}}{\mathbf{a} \cdot (\mathbf{b} \times \mathbf{c})}; \quad \mathbf{c}^* = \frac{\mathbf{a} \times \mathbf{b}}{\mathbf{a} \cdot (\mathbf{b} \times \mathbf{c})}. \qquad (2.10)$$

The new co-ordinate axes are perpendicular respectively to the three old co-ordinate planes. The lattice generated by the three vectors $\mathbf{a}^*, \mathbf{b}^*, \mathbf{c}^*$ is called the *reciprocal lattice* of the lattice generated by $\mathbf{a}, \mathbf{b}, \mathbf{c}$. It is clear from (2.9) that this is a reciprocal relationship; in other words, the reciprocal of the reciprocal lattice is the original lattice.

It often happens that the same idea arises independently in different places, and is then given different names. Indeed, relations of the type (2.9) occur in transformation theory, relativity theory, etc. In that case the vectors $\mathbf{a}, \mathbf{b}, \mathbf{c}$ are referred to as *covariant*, the vectors $\mathbf{a}^*, \mathbf{b}^*, \mathbf{c}^*$ as *contravariant* or vice versa. The reason for this name is that if a matrix $\mathfrak{M}$ defines a transformation in the covariant system, the same matrix, with an interchange of lines and columns, defines the inverse transformation in the contravariant system. In the case of the simple cubic lattice, the original lattice and the reciprocal one differ from each other only in the scale. However, this is only true for that lattice. The reciprocal lattice of the body-centred cubic lattice is face-centred cubic, as can easily be verified. In general, the reciprocal lattice is not of the same type as the original Bravais lattice. Its point symmetries must be the same however, because the definition (2.9) is intrinsically symmetric. Hence the reciprocal lattice always belongs to the same crystal system.

The advantage of reciprocal space for X-ray theory is that its introduction interchanges the properties of planes and points. We shall check up on this notion only as far as we need it, namely, by analysing a lattice vector $\mathbf{r}^*$ of the reciprocal lattice which we define in analogy to (1.2) by the relation

$$\mathbf{r}^*_{LMN} = L\,\mathbf{a}^* + M\,\mathbf{b}^* + N\,\mathbf{c}^*, \qquad (2.11)$$

where L, M, N are integers. This vector has the following two properties:

(a) $\mathbf{r}^*_{LMN}$ *is perpendicular to the lattice planes* (LMN) *or* $(L/\sigma, M/\sigma, N/\sigma)$.

To prove this let
$$\mathbf{r} = \alpha\,\mathbf{a} + \beta\,\mathbf{b} + \gamma\,\mathbf{c}$$

be a vector in the (LMN) plane passing through the origin. Its components then obey a special form of the equation (1.6), namely

$$L\alpha + M\beta + N\gamma = 0.$$

Now take the dot-product of $\mathbf{r}^*_{LMN}$ and $\mathbf{r}$

$$\mathbf{r}^*_{LMN} \cdot \mathbf{r} = (L\mathbf{a}^* + M\mathbf{b}^* + N\mathbf{c}^*) \cdot (\alpha\,\mathbf{a} + \beta\,\mathbf{b} + \gamma\,\mathbf{c})$$
$$= L\alpha + M\beta + N\gamma = 0$$

because of (2.9).

(b) *If* $\mathbf{r}^*_{LMN}$ *is the shortest lattice vector in its own direction, then its length equals the reciprocal of the distance between the associated lattice planes.*

For proof, we observe that L, M, N by assumption, have no common divisor except ± 1. Now place a point P_1 of co-ordinates

x_1, y_1, z_1 on one lattice plane and P_2 with x_2, y_2, z_2 on the adjacent one. Then

$$Lx_1 + My_1 + Nz_1 = D,$$
$$Lx_2 + My_2 + Nz_2 = D+1,$$
$$L(x_2 - x_1) + M(y_2 - y_1) + N(z_2 - z_1) = 1.$$

Now take the scalar product of $\mathbf{r}_2 - \mathbf{r}_1$ with $\mathbf{r}^*_{LMN}$

$$\mathbf{r}^*_{LMN} \cdot (\mathbf{r}_2 - \mathbf{r}_1) =$$
$$= (L\mathbf{a}^* + M\mathbf{b}^* + N\mathbf{c}^*) \cdot [(x_2 - x_1)\mathbf{a} + (y_2 - y_1)\mathbf{b} + (z_2 - z_1)\mathbf{c}]$$
$$= L(x_2 - x_1) + M(y_2 - y_1) + N(z_2 - z_1) = 1;$$

that is, the scalar product of the two vectors is unity. If $\mathbf{r}_2 - \mathbf{r}_1$ is chosen parallel to $\mathbf{r}^*_{LMN}$, it is perpendicular to the lattice planes, and its magnitude equals their distance. At the same time the scalar product becomes an ordinary product. Hence the length of $\mathbf{r}^*_{LMN}$ equals the reciprocal distance.

After these preliminaries, it is perhaps not too surprising that the concept of the reciprocal lattice permits explicit solution of the equations (2.6). Indeed, if we express $\mathbf{k} - \mathbf{k}_0$ in terms of the basis vectors of that lattice

$$\mathbf{k} - \mathbf{k}_0 = L^*\mathbf{a}^* + M^*\mathbf{b}^* + N^*\mathbf{c}^*,$$

then we get immediately from (2.6) and (2.9)

$$L^* = L, \quad M^* = M, \quad N^* = N.$$

In other words, with (2.1)

$$\mathbf{k} - \mathbf{k}_0 = \mathbf{r}^*_{LMN}. \quad (2.12)$$

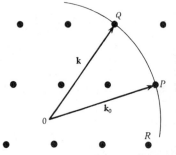

Equation (2.12) is made use of in the construction of Bragg reflections which was introduced by Ewald.[†] In this construction one takes the reciprocal lattice and places the wave vector of the incoming radiation $\mathbf{k}_0$ so as to terminate on a lattice point P (fig. 2.3). One then draws a sphere around the starting point O of this vector

Fig. 2.3. Laue–Bragg scattering viewed by the Ewald method; Ewald sphere construction in the reciprocal lattice.

(Ewald sphere) which is made to pass through P. If it passes

† P. P. Ewald, Z. Kristallogr. **56** (1921), 129.

through any other lattice points besides, such as Q, then the vector OQ is the wave vector of a possible reflected Bragg wave. The method can be extended to answer questions such as whether the lattice point R, which is close to the sphere, produces a reflection. In order to do this an extension must be associated with each lattice point and perhaps also a thickness with the sphere. This leads us into questions of line width which, in the simplest cases, can be handled in the way we treated the diffraction by our crystal block (diffraction by mosaic blocks). Many cases are more complicated, however. No attempt will be made to treat this subject here.

In conclusion, a word may be added anticipating the use of the reciprocal lattice in other fields of solid state physics. We see from comparison of (2.12) and (2.6) that if the wave vectors of two waves are made to differ from each other by a vector of the reciprocal lattice, their phase difference between corresponding points in the crystal lattice will be equal. This is most easily visualized by expanding (2.6) to read, apart from an integer,

$$\mathbf{k}\cdot(\mathbf{x}+\mathbf{r})-\mathbf{k}\cdot\mathbf{x}=\mathbf{k_0}\cdot(\mathbf{x}+\mathbf{r})-\mathbf{k_0}\cdot\mathbf{x},$$

where $\mathbf{x}$ is an arbitrary position vector and $\mathbf{r}$ is a vector of the Bravais lattice obeying (1.2). We can therefore say that from the point of view of the Bravais lattice, the two waves are equivalent, because although they differ from each other within a primitive cell, they behave equivalently from cell to cell. X-ray diffraction, in this view, is then due to the inability of the lattice to distinguish between two such waves. This is a little far fetched for X-rays which are electromagnetic waves existing independently of the lattice, in the 'ether'. In later applications, however, the waves will be waves of the material making up the lattice. Then it is indeed no longer possible to distinguish between $\mathbf{k}$ and $\mathbf{k_0}$. For such applications wave vectors are preferably reduced to their smallest possible value which can be realized by subtraction of the nearest vector of the reciprocal lattice. The wave vector so obtained is called the *reduced wave vector*. The terminal points of all reduced wave vectors fill a small region around the origin of reciprocal space, which is the Wigner–Seitz primitive cell of the reciprocal lattice. This cell is given the name *First Brillouin Zone*.

PROBLEMS

1. Evaluate and plot the expression (2.5) as a function of x for $L = 2$, 4, 8. Comment on the result.

2. Show that if a matrix $\mathfrak{M}$ effects the transformation of three vectors a, b, c into three new ones, the transpose of $\mathfrak{M}$ (lines and columns exchanged) effects the inverse transformation among the reciprocal vectors a*, b*, c* associated with a, b, c.

3. Show that the reciprocal lattice of the body-centred cubic lattice is face-centred cubic.

4. Prove that if two lattices are reciprocals of each other the product of their primitive cell volumes equals unity.

5. Derive the expression analogous to (2.8) for a crystal whose Bravais lattice is face-centred with respect to the indexing cell adopted. What are the missing reflections for such a crystal?

6. Utilize the results of problem 10 of the previous chapter to index the reflections of a rhombohedral crystal on a hexagonal cell and vice versa. Derive from this the missing reflections; check on the result by using (2.3).

7. Beyond reflections missing by symmetry, formula (2.3) predicts 'forbidden lines' for many structures because identical atoms occur in non-equivalent crystalline positions. Use this method to find the forbidden lines in the close-packed hexagonal, tellurium and diamond structures.

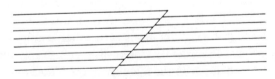

Fig. 2.4. Schematic representation of a crystalline film having two parts shifted by a fraction of a lattice distance.

8. In the adjoining fig. 2.4 there is shown a schematic drawing of a crystalline film. The right-hand side of the film is shifted in depth with respect to the left by a fraction of a lattice distance. The border between the two portions does not run perpendicular to the film. Show with the help of equation (2.3) that if the film is examined with the help of normally incident electrons then the strip over which the two regions overlap is recognizable by interference fringes, provided the normally transmitted beam is close to obeying the Bragg condition.†

† M. J. Whelan, P. B. Hirsch, R. W. Horne and W. Bollmann, *Proc. Roy. Soc.* A, **240** (1957), 524.

BIBLIOGRAPHY

X-RAY DIFFRACTION

G. L. CLARK. *Applied X-rays*, 4th ed. (New York, McGraw-Hill, 1955.)
R. W. JAMES. *The Optical Principals of the Diffraction of X-rays.* (London, Bell and Son, 1948.)
W. H. ZACHARIASEN. *Theory of X-ray Diffraction in Crystals.* (New York, Wiley, 1945.)

ELECTRON DIFFRACTION

Z. G. PINSKER. *Electron Diffraction.* (Transl.) (London, Butterworths, 1953.)
G. P. THOMSON and W. COCHRANE. *Theory and Practice of Electron Diffraction.* (London, Macmillan, 1939.)

NEUTRON DIFFRACTION

G. E. BACON. *Neutron Diffraction.* (Oxford, Clarendon Press, 1955.)
C. G. SHULL and E. O. WOLLAN. Applications of neutron diffraction to solid state problems. *Solid State Physics*, 2 (1956), 137–217.

CHAPTER 3

DYNAMICS OF LATTICE VIBRATIONS

The geometrical discussion of the last two chapters implies that the solid is in static equilibrium. This is true only in a limited sense. First of all, it is not at all true for electrons which are, just as in atoms or molecules, in stationary states of motion involving energies of the order of electron volts. This radical departure from statics will be discussed later. A less serious restriction applies to the atomic nuclei; their positions are essentially fixed, but small departures from equilibrium are possible. These departures will be the subject of the present discussion.

Whenever a system of particles is almost in static equilibrium their motion may be analysed by vibration theory, a well understood special chapter of mechanics.† In that discipline the potential energy of the system is developed in powers of the particle displacements from equilibrium. If this development is broken off at quadratic terms (which is legitimate for small enough amplitudes of vibration) the potential energy appears as a homogeneous quadratic form, the linear terms being absent because of the equilibrium conditions. General theory then goes on to prove that there exist so-called *normal co-ordinates*, which are linear functions of the displacements. In terms of these normal co-ordinates X_i both the kinetic energy K and the potential energy U are expressed as a sum of squares

$$K = \tfrac{1}{2} \sum_i \left(\frac{\mathrm{d}X_i}{\mathrm{d}t}\right)^2, \quad U = 2\pi^2 \sum_i f_i^2 X_i^2. \tag{3.1}$$

The mechanical system is thus equivalent to a set of uncoupled harmonic oscillators whose frequencies are f_i.

We shall apply in this chapter the general vibration theory to vibrations of crystals. We shall find that the ordered structure of the medium produces an essential simplification: the normal

† See for instance E. T. Whittaker, *Analytical Dynamics* (Cambridge, University Press, 1937), Chapter VII, or H. Goldstein, *Classical Mechanics* (Cambridge, Mass., Addison-Wesley Press, 1951, Chapter 10).

co-ordinates are amplitudes of plane elastic waves traversing the crystal. The oscillator quanta going with those waves are known as *phonons*.

SIMPLE MECHANICAL MODELS

As in many other fields of physics, the understanding of lattice vibrations in general is greatly aided by the discussion of model cases capable of explicit solution. The aspect which simplifies the subject most markedly is the fact that the vibrating medium is periodic in space. A simple model of such a medium consists of a linear array of equal masses M connected by equal massless springs of spring constant k (fig. 3.1); let us investigate the possible

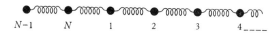

$N-1$ N 1   2   3   4 ____

Fig. 3.1. One-dimensional array of equal masses and springs; this is the simplest model of a vibrational band.

longitudinal motions of such an array. For simplicity, the first mass is assumed linked up with the last one. This is entirely to avoid special consideration of the end members. It stands to reason that the disposal of the end members has no profound influence upon the modes of motion if the number N of masses is assumed large. The problem is thus one of N degrees of freedom, with co-ordinates $x_0, x_1, x_2, x_3, ..., x_{N-1}$, which measure the displacement of each mass from its equilibrium position. It is sometimes convenient to look at this same problem as if the array were infinite, but subject to a periodic boundary condition†

$$x_{n+N} = x_n. \qquad (3.2)$$

The kinetic energy K of the system so defined equals

$$K = \tfrac{1}{2} M \sum_{\nu=0}^{N-1} \left(\frac{dx_\nu}{dt}\right)^2, \qquad (3.3)$$

and the potential energy

$$U = \tfrac{1}{2} k \sum_{\nu=0}^{N-1} (x_\nu - x_{\nu+1})^2. \qquad (3.4)$$

† This boundary condition is sometimes called the Born–von Karman boundary condition.

The classical equations of motion derived from this are

$$M\frac{d^2x_n}{dt^2} = k\left(x_{n+1} + x_{n-1} - 2x_n\right).\tag{3.5}$$

The equations are of such a form that normal mode solutions are assured in principle, but one might well expect that finding them would be difficult. We shall see below how the translational symmetry of the crystal permits us to find those modes a priori, regardless of the couplings within the crystal. We shall anticipate, for the moment, on this result and employ the 'trick' of multiplying (3.5) with

$$e^{2\pi i p n/N},$$

where p is an integer, and summing over n. The result is conveniently expressed in terms of a co-ordinate X_p which we define as

$$X_p = \frac{1}{\sqrt{N}} \sum_{\nu=0}^{N-1} e^{2\pi i p \nu/N} x_\nu.\tag{3.6}$$

In terms of this variable, we find

$$\frac{d^2X_p}{dt^2} + 4\pi^2 f_p^2 X_p = 0,\tag{3.7}$$

where the frequency f_p equals

$$f_p = \frac{1}{\pi}\left(\frac{k}{M}\right)^{\frac{1}{2}} \sin\frac{\pi|p|}{N}.\tag{3.8}$$

In interpreting these results we may start by observing that the number of co-ordinates X_p is exactly equal to N. First, there are not more than N; for, if p is replaced by $p + N$ in (3.6) the defining equation for X_p remains identical to itself. Second, there are not less; for the N remaining variables are linearly independent, as is proved by the possibility of solving (3.6) with respect to x_n. This is done by multiplication of (3.6) by

$$e^{-2\pi i p n/N}$$

and summation over p. Because of the law of summation for geometric series this procedure isolates a single x_n and yields:

$$x_n = \frac{1}{\sqrt{N}} \sum_{\mu=0}^{N-1} e^{-2\pi i \mu n/N} X_\mu.\tag{3.9}$$

This equation also exhibits the nature of the variable X_p. If we assume all variables X_μ to be zero but one, and if we set for that one from (3.7)

$$X_p(t) = X_p(0) e^{2\pi i f_p t}, \qquad (3.10)$$

we get

$$x_n(t) = \frac{1}{\sqrt{N}} X_p(0) \, \mathrm{ixp} \left[f_p t - \frac{pn}{N} \right], \qquad (3.11)$$

where for convenience the abbreviation

$$\mathrm{ixp}\, \mu = e^{2\pi i \mu} \qquad (3.12)$$

has been introduced. We see that the number p appears here as proportional to the wave number of a running wave. This number, for a wave through a lattice of discrete points, is only specified up to a reciprocal lattice vector; this was already discussed at the end of the previous chapter. It is reasonable to remove this ambiguity by placing p into the 'first Brillouin zone', that is to set

$$-\tfrac{1}{2}N < p \leqq +\tfrac{1}{2}N. \qquad (3.13)$$

We also see that the variables X_p are not yet the normal modes, because the displacement resulting from one of them is complex. It follows in fact from (3.9) that the restriction

$$X_{-p} = X_p^* \qquad (3.14)$$

must hold. This means that X_0 and $X_{\frac{1}{2}N}$ (if N is even) must be real, and that independent normal co-ordinates can be defined by the prescription

$$X_p = \frac{1}{\sqrt{2}} (A_p + i B_p), \qquad (3.15a)$$

$$X_{-p} = \frac{1}{\sqrt{2}} (A_p - i B_p), \qquad (3.15b)$$

$$A_p, B_p \text{ real}, \qquad (3.15c)$$

$$0 < p < \tfrac{1}{2}N. \qquad (3.15d)$$

The quantities A_p, B_p obey the same uncoupled oscillator equations (3.7) as the quantities X_p. From these equations, and (3.9) and (3.15), the displacements due to one single A_p come out to be

$$x_n(t) = \left(\frac{2}{N} \right)^{\frac{1}{2}} A_p(0) \cos \frac{2\pi p n}{N} \cos 2\pi f_p (t - t_0). \qquad (3.16)$$

For the B_p's the first trigonometric function is a sine. Equation (3.16) represents a standing wave of wave number

$$k = \frac{p}{Nd},\qquad(3.17)$$

where d is the distance between neighbouring masses. Running waves travelling either to the right or the left can be built up from them by forming proper linear combinations. A simplified viewpoint which is often useful is to observe that running waves are implicitly present in (3.11), if one considers separately its real and imaginary

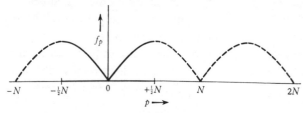

Fig. 3.2. Plot of frequency versus wave number for the model of fig. 3.1; all relevant information is contained in the part shown in full outline.

parts. Either of these is a solution of the basic equations, since the equations are real. In this sense the quantities X_p themselves can be treated as normal modes. This viewpoint will be adopted later occasionally as a convenient shorthand procedure.

With the normal modes X_0, $X_{\frac{1}{2}N}$, A_p, B_p goes a spectrum of frequencies f_p given by (3.8). The case $p = 0$ is a limiting one of zero frequency; it shifts the entire chain uniformly and has no restoring force. The other limiting case is $p = \frac{1}{2}N$; it corresponds to alternate masses moving in opposite directions and has the maximum frequency

$$f_{\max} = \frac{1}{\pi}\left(\frac{k}{M}\right)^{\frac{1}{2}}.\qquad(3.18)$$

The remainder of the frequencies forms a quasicontinuous band between these two limits. Each frequency goes with two modes, which are waves travelling in opposite directions. Because of the indeterminacy in the wave number, the relation giving f_p as a function of p must be periodic in p and must be completely described if it is known in the first Brillouin zone. Indeed (3.8) has the required periodicity. The relation is plotted in fig. 3.2; the part falling into the range (3.13) is shown in full outline, and the periodic continuation of it is shown dashed. A feature worth noting about this figure is the sharp corner in the frequency versus wave-number curve near the origin; this produces a linear relation between the frequency and the absolute value of the wave number, provided both are small. The wave length formula (3.17) and the limit of (3.8) for low p/N thus yield a wave phenomenon whose fixed velocity of propagation equals

$$c = \left(\frac{k}{M}\right)^{\frac{1}{2}} d. \tag{3.19}$$

These limiting considerations establish the connexion between our model and the theory of elastic waves in a continuum.

The application of quantum theory to the model of fig. 3.1 is almost trivial. In order to bring the problem into Hamiltonian form we substitute (3.9) into (3.3) and (3.4). The result is

$$K = \frac{1}{2} M \sum_{\mu=0}^{N-1} \frac{dX_\mu}{dt} \frac{dX_{-\mu}}{dt}, \tag{3.20a}$$

and from (3.8)

$$U = 2\pi^2 M \sum_{\mu=0}^{N-1} f_\mu^2 X_\mu X_{-\mu}. \tag{3.20b}$$

As observed previously, the variables X_μ are not yet the normal co-ordinates; indeed, the relations (3.20) still differ essentially from (3.1). Further substitution of (3.15) yields finally

$$K = \frac{1}{2} M \left(\frac{dX_0}{dt}\right)^2 + \frac{1}{2} M \left(\frac{dX_{\frac{1}{2}N}}{dt}\right)^2 + \frac{1}{2} M \sum_{\mu=1}^{[\frac{1}{2}N]} \left[\left(\frac{dA_\mu}{dt}\right)^2 + \left(\frac{dB_\mu}{dt}\right)^2\right], \tag{3.21}$$

$$U = 2\pi^2 M f_{\max}^2 X_{\frac{1}{2}N} + 2\pi^2 M \sum_{\mu=1}^{[\frac{1}{2}N]} f_\mu^2 [A_\mu^2 + B_\mu^2]. \tag{3.22}$$

Here the terms in $X_{\frac{1}{2}N}$ are absent if N is odd, and $[\frac{1}{2}N]$ is the largest integer smaller than $\frac{1}{2}N$. Apart from an unimportant mass factor these expressions have the form (3.1). Their quantization is immediate. The degree of freedom corresponding to X_0 is not quantized; the others are quantized in quanta of magnitude

$$E_\mu = h f_\mu, \tag{3.23}$$

with half that amount for zero point energy.

Let us now modify the preceding example by making the even and odd masses different, but leaving everything else equal. The total number of masses must then be even and will be denoted by $2N$. N is then the number of 'primitive cells' of the array. There are now two masses per cell. The equations of motion are

$$
\left.
\begin{aligned}
M\frac{d^2 x_{2n}}{dt^2} &= k(x_{2n+1}+x_{2n-1}-2x_{2n}), \\
m\frac{d^2 x_{2n+1}}{dt^2} &= k(x_{2n+2}+x_{2n}-2x_{2n+1}),
\end{aligned}
\right\}
\tag{3.24}
$$

with the boundary conditions

$$
x_{k+2N} = x_k. \tag{3.25}
$$

We may now take for granted that each normal mode is a sinusoidal wave, whose wave number is specified by (3.17). The only point to watch is that the M-wave and the m-wave may have different amplitudes and phases. We do this by setting instead of (3.9)

$$
\left.
\begin{aligned}
x_{2n} &= \frac{1}{\sqrt{N}}X_p\,e^{-\pi i 2np/N}, \\
x_{2n+1} &= \frac{1}{\sqrt{N}}Y_p\,e^{-\pi i(2n+1)p/N}.
\end{aligned}
\right\}
\tag{3.26}
$$

The choice of X_p and Y_p is made so that their phase difference measures the phase difference of the M-wave and the m-wave at the same point in space. To make this definite we must choose p according to the prescription (3.13). This makes the phase difference between consecutive M-masses less than π and assigns to the sandwiched m-mass a formal phase equal to their arithmetic mean. The actual phase is measured as the departure from this value. Substitution of (3.26) into (3.24) yields

$$
M\frac{d^2 X_p}{dt^2} = -2kX_p+2kY_p\cos\frac{\pi p}{N},
$$

$$
m\frac{d^2 Y_p}{dt^2} = -2kY_p+2kX_p\cos\frac{\pi p}{N}.
$$

The equation system allows two circular frequencies ω, which follow from the secular equation

$$
\begin{vmatrix}
2k-M\omega^2 & -2k\cos\dfrac{\pi p}{N} \\[2ex]
-2k\cos\dfrac{\pi p}{N} & 2k-m\omega^2
\end{vmatrix}
= 0,
$$

or $\quad \omega^2 = \dfrac{k(m+M) \pm [k^2(M+m)^2 - 4k^2 Mm \sin^2 \pi p/N]^{\frac{1}{2}}}{Mm}$, $\qquad$ (3.27)

and finally $\quad \dfrac{Y_p}{X_p} = \dfrac{2k - M\omega^2}{2k \cos \pi p/N} = \dfrac{2k \cos \pi p/N}{2k - m\omega^2}$. $\qquad$ (3.28)

From consultation of (3.26), (3.27) and (3.28), the reader can easily verify the following facts:

(a) $p = 0$ yields both the lowest and the highest frequency. The lowest frequency is zero and all masses move in unison. The highest frequency is given by

$$\omega_{\max}^2 = 2k\left(\frac{1}{M} + \frac{1}{m}\right).$$

In this set, all M-masses move in unison against all m-masses with amplitudes inversely as the masses.

(b) $p = \frac{1}{2}N$ yields the lowest frequency ω_L arising from the positive value of the square root in (3.27); it equals

$$\omega_L^2 = \frac{2k}{m}.$$

In this mode the large M-masses stand still, and the m-masses move alternately in opposite directions. It also yields the highest frequency for negative square root which is

$$\omega_H^2 = \frac{2k}{M},$$

and where the role of the masses is reversed. Since we have

$$\omega_L^2 > \omega_H^2$$

the frequencies with the two signs of the square root are separated by a gap, which increases as the masses become more unequal.

These results establish the existence of two branches, separated by a gap, as exhibited in fig. 3.3. These branches differ qualitatively from each other, because the phase difference between the M-wave and the m-wave is π for the upper one and 0 for the lower. To see this, we observe first that it is always either 0 or π because the ratios (3.28) are real. To find which it is observe from (3.28) that

$$2\frac{k}{m} - \omega^2 \text{ and } 2\frac{k}{M} - \omega^2$$

have the same sign, which is the sign of Y_p/X_p. This sign must be also the sign of their arithmetic mean

$$k\left(\frac{1}{m}+\frac{1}{M}\right)-\omega^2,$$

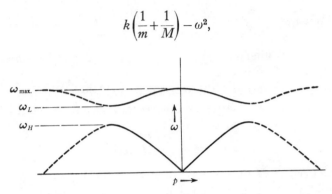

Fig. 3.3. Plot of frequency versus wave number for a one-dimensional model with two alternate masses; there are now two frequency bands.

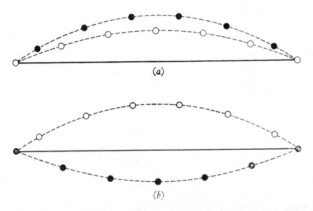

Fig. 3.4. Nature of the vibration in the acoustical (a) and optical (b) branch of a vibration spectrum; the open and full circles represent different atoms.

which by (3.27) equals the negative of the sign of the square root. This proves the statement. An exception must be allowed for the endpoints ω_L and ω_H, where the assignment of a phase difference becomes meaningless because one of the amplitudes is zero. It is customary to refer to the two branches of the frequency spectrum as the *acoustical* and *optical* branch respectively. The name derives from the case of the alkali halides, of which this example is a simple model. In the acoustical branch the two types of ions move in phase, in the optical branch out of phase. This state of affairs is qualitatively exhibited in fig. 3.4. When the two kinds of masses also represent opposite charges then the second type of motion will couple very strongly to the electromagnetic field. Indeed, all the alkali halides have a strong band in the infra-red, the so-called 'reststrahl' band, which is of this nature. By analogy, the term 'optical branch' is used even in crystals like germanium, which has two atoms per primitive cell and hence two branches in its vibration spectrum.

REPRESENTATION THEORY OF THE CRYSTAL TRANSLATION
GROUP

In passing from these models to the vibrations of actual crystals it will be useful to elucidate first the group theoretical background which enabled us to pick out the normal modes with so little labour.

The theory of *matrix representation of groups* occupies itself with the question what properties a series of quantities $X_1, X_2, ..., X_d$ have which all obey the same homogeneous linear equation having certain symmetries. In the present case $X_1 ... X_d$ are a complete linearly independent set of co-ordinates having a fixed frequency f. If the symmetry operation of the basic equation is applied to one of these X's it will generate another co-ordinate going with this frequency; this co-ordinate is necessarily a linear combination of the original co-ordinates if the set was complete to start with.

To make these ideas clear the reader may think of the normal modes of vibration of a simple molecule as an example. Fig. 3.5 shows two normal modes of vibration of the molecule of ammonia. The motions of the three symmetrical hydrogen atoms are shown with respect to the nitrogen atom, which is taken as fixed for convenience. Each of these normal modes of vibration is representable by a co-ordinate X_i. Now as we rotate either figure (a) or (b) by 120° a new figure is produced. The rotation performed is a symmetry operation for the molecule, and therefore, the new figure corresponds again to a normal mode whose frequency equals the original one. Now it so happens that in fig. 3.5 (a) the new mode equals the original one, while in fig. 3.5 (b) it does not. Thus, if we represent the rotational operation by $\mathfrak{S}$, the mode 3.5 (a) by X_1 and the mode 3.5 (b) by X_2 we may write

$$\mathfrak{S}X_1 = X_1,$$

$$\mathfrak{S}X_2 = X_3,$$

where X_3 is a new mode to be added to the first two. Quite generally, the result of operating with a symmetry operation $\mathfrak{S}$ on a set of normal co-ordinates X_i will be a linear relation of the form

$$\left.\begin{aligned}
\mathfrak{S}X_1 &= a_{11}X_1 + a_{12}X_2 + \ldots + a_{1d}X_d, \\
\mathfrak{S}X_2 &= a_{21}X_1 + a_{22}X_2 + \ldots + a_{2d}X_d, \\
&\ldots\ldots\ldots\ldots\ldots\ldots\ldots\ldots\ldots\ldots\ldots\ldots\ldots\ldots\ldots \\
\mathfrak{S}X_d &= a_{d1}X_1 + \ldots\ldots\ldots + a_{dd}X_d.
\end{aligned}\right\} \qquad (3.29)$$

We see that through the equation system (3.29) a matrix of dimension d is associated with each element of the symmetry group. These

Fig. 3.5. Two normal modes of vibration of the ammonia molecule. The mode (a) is non-degenerate, while the mode (b) is degenerate for symmetry reasons.

matrices are said to form a *representation* of the group, because successive operation with group elements (group multiplication) corresponds to successive multiplication by the associated matrices, and hence, by the associative law, multiplication by the matrix product. It follows that the matrix product must be the associate of the group product; therefore, the abstract rules of group multiplication are realized by a set of matrices, which form their representation. In this sense, the matrices in (1.1) *represent* the rotation-reflection group in three-dimensional space; the representation is furnished through the intermediary of the cartesian co-ordinates x, y, z, whose transformation generates the matrix representation. Generally the set X_1, X_2, ..., X_d which generates the representation is called the *representation module* of the group; d is its *dimension*. The basic question of representation theory is whether this number d can be *reduced*, that is whether suitable linear combinations of the X's can be found such that a subset module of less than d dimensions transforms within itself under the operations of the group. We may return to our example of NH_3 to illustrate this point. To represent the internal degrees of freedom of the molecule we need $3 \times 4 - 3 - 3 = 6$ co-ordinates. For these co-ordinates the $120°$ rotation $\mathfrak{S}$ is represented by a 6 by 6 matrix, in other words, the representation module is six-dimensional. The question of reducibility is answered here by fig. 3.5 (a); the co-ordinate represented in this figure is invariant under all symmetry operations of the molecule and thus forms by itself a subset module of dimension 1. In the case of fig. 3.5 (b), on the other hand, the dimension of the representation is at least 2 as shown above; careful analysis shows that it is in fact exactly 2 because the result of operating with $\mathfrak{S}$ on X_3 produces only a linear combination of X_2 and X_3, and not a new co-ordinate. The theory of representation of groups permits us to make a priori statements about such representations by considering the symmetry group only and ignoring the basic equations of which the quantities X_i are solutions. The result of such a study is a table of all possible fully reduced representations of a given symmetry group, and criteria which permit us to find these representations within a module which is not yet fully reduced (such as the six internal co-ordinates of ammonia discussed above). It is seen from the example of fig. 3.5 (a) that such a reduction might lead in an extreme case to quantities X_i which are multiplied by a factor under all operations of the group. The representation module of dimension d is then said to be reduced to a sum of d modules of one dimension, and the representation is *completely reduced* to a sum of one-dimensional representations. We cannot expect, however; that a reduction to one dimension is generally possible; for most groups (such as the rotation-reflection group discussed at the beginning of Chapter 1) do not obey the commutative law of multiplication, that is $AB \neq BA$; neither, usually do the representing matrices such as the ones written out in (1.1). On the other hand, a sum of one dimensional representations is commutative, as the reader may verify directly, and therefore the general reduction of all matrices to this form cannot be a possibility. But for the translational subgroup of the total crystalline space group this obvious objection is not valid since the subgroup is commutative. Indeed we shall now prove the following theorem:

In the field of complex numbers any representation of a crystal translation group can be reduced to a sum of one dimensional representations.

For ease of reasoning, assume the Bravais lattice finite with periodic boundary conditions: let L, M, N be the periods along **a**, **b**, **c**. Let x be a member of a representation module. The result of operating on x with the crystal translation operator $(l\mathbf{a} + m\mathbf{b} + n\mathbf{c})$ is then defined, and within the same module; we denote it by $x(-l, -m, -n)$. The essence of our proof is now that certain roots of unity do furnish representations of the translation group and that they are sufficiently numerous to allow no other representation. Indeed, let us develop $x(l, m, n)$ into a finite trigonometric series. With the help of the abbreviation (3.12) we may give to this series the form

$$x(l, m, n) = \frac{1}{\sqrt{LMN}} \sum_{l^*=0}^{L-1} \sum_{m^*=0}^{M-1} \sum_{n^*=0}^{N-1}$$

$$\mathrm{ixp}\left(-\frac{ll^*}{L} - \frac{mm^*}{M} - \frac{nn^*}{N} \right) X(l^*, m^*, n^*), \quad (3.30)$$

in obvious analogy to (3.9). An expansion of this type is always possible for arbitrary $x(l, m, n)$; for it is seen from (3.6) and (3.9) that this type of series can be inverted. In the present case, we multiply (3.30) with

$$\mathrm{ixp}\left(\frac{ll^+}{L} + \frac{mm^+}{M} + \frac{nn^+}{N} \right),$$

where l^+, m^+, n^+ are integers; thereupon, we sum over l, m, n, the summation being over L, M, N successive integers. The result is

$$X(l^+, m^+; n^+) = \frac{1}{\sqrt{LMN}} \sum_{l=0}^{L-1} \sum_{m=0}^{M-1} \sum_{n=0}^{N-1}$$

$$\mathrm{ixp}\left(\frac{ll^+}{L} + \frac{mm^+}{M} + \frac{nn^+}{N} \right) x(l, m, n). \quad (3.31)$$

Because of (3.31) the X's are members of the representation module generated by x, and because of (3.30) any member of the module is expressible in terms of them. Each quantity $X(l^+, m^+, n^+)$ is, however, either equal to zero, or else by itself a one-dimensional representation module. To show this, operate on (3.31) with the translation $\mathfrak{T} = (l'\mathbf{a} + m'\mathbf{b} + n'\mathbf{c})$; because of the associative nature of successive translations, this yields

$$\mathfrak{T}X(l^+, m^+, n^+) =$$

$$= \frac{1}{\sqrt{LMN}} \sum_l \sum_m \sum_n \mathrm{ixp} \left(\frac{ll^+}{L} + \frac{mm^+}{M} + \frac{nn^+}{N} \right) x(l - l', m - m', n - n')$$

$$= \frac{1}{\sqrt{LMN}} \sum_{l,m,n} \mathrm{ixp} \left(\frac{(l+l')l^+}{L} + \frac{(m+m')m^+}{M} + \frac{(n+n')n^+}{N} \right) \times$$

$$\times x(l, m, n),$$

or finally

$$\mathfrak{T}X(l^+, m^+, n^+) = \mathrm{ixp} \left(\frac{l'l^+}{L} + \frac{m'm^+}{M} + \frac{n'n^+}{N} \right) \times X(l^+, m^+, n^+).$$

$$(3.32)$$

This proves the theorem. Equation (3.32) exhibits, incidentally, all possible values which the representing multiplier of the group element $(l'\mathbf{a} + m'\mathbf{b} + n'\mathbf{c})$ can take. There are LMN different representations, corresponding to the LMN *effectively* different possible values of the parameters l^+, m^+, n^+.

This powerful theorem, and the proof given, can be made to cover all groups for which the commutative law of multiplication holds (Abelian groups). On the other hand, restriction to matrices with elements in the field of complex numbers is essential. The difficulty with the X's in the present instance is exactly here. The normal co-ordinates must be such that they furnish a representation of the crystal translation group in the *real* number field. This is true for at most two of the X's defined by (3.6), and hence for at most eight of the representations (3.32). In all other cases the definition of the quantity $X(l^+, m^+, n^+)$ must be followed, because of (3.14), by the transformation (3.15). The two co-ordinates $A(l^+, m^+, n^+)$ and $B(l^+, m^+, n^+)$ then form together a module which furnishes a two-dimensional real matrix representation of the crystal translation group, which is irreducible in the field of real numbers.

VIBRATIONS OF THREE-DIMENSIONAL LATTICES

Because of the group theoretical foreknowledge gained we can approach the general case of vibrating crystals with relative confidence. Group theory furnishes us a priori with all possible modes and their wave-lengths. It remains for dynamics to sort out branches (such as optical and acoustical), if present, and to determine polarizations. In the simplest case we may assume a number of identical atoms located on a simple Bravais lattice with L, M, N atoms along the three fundamental directions. The number of dynamical variables of the system will then be $3LMN$. Let the problem of the surface be eliminated by periodic boundary conditions, as previously. Then the equation system will, to start with, have the following form

$$\frac{\mathrm{d}^2 x^i_{lmn}}{\mathrm{d}t^2} = \sum_{\lambda=0}^{L-1} \sum_{\mu=0}^{M-1} \sum_{\nu=0}^{N-1} \sum_{k=1}^{3} \kappa^{ik}_{\lambda\mu\nu} x^k_{l+\lambda,\,m+\mu,\,n+\nu}, \qquad (3.33\,a)$$

where

$$\left.\begin{aligned} i &= 1, 2, 3, \\ l &= 0, 1, 2, 3, ..., L-1, \\ m &= 0, 1, 2, 3, ..., M-1, \\ n &= 0, 1, 2, 3, ..., N-1. \end{aligned}\right\} \qquad (3.33\,b)$$

i and k label the three co-ordinate axes; l, m, n, λ, μ, ν label the three cell numbers of each primtive cell. $\kappa^{ik}_{\lambda\mu\nu}$ gives the elastic coupling (divided by a mass) of the i-co-ordinate of the atom l, m, n to the k-co-ordinate of the atom $l+\lambda$, $m+\mu$, $n+\nu$. The crystalline symmetry enters into the expression because the coupling constant depends only on the *difference* of the cell numbers, not on the cell numbers themselves. This feature permits introduction of plane waves as normal co-ordinates. Applying (3.31) we introduce

$$X^i_{l^*,\,m^*,\,n^*} = \frac{1}{\sqrt{LMN}} \sum_{l,\,m,\,n} \mathrm{i} \mathrm{x} \mathrm{p} \left(\frac{ll^*}{L} + \frac{mm^*}{M} + \frac{nn^*}{N}\right) x^i_{l,\,m,\,n}. \qquad (3.34)$$

The equation of motion for this co-ordinate results by multiplication of (3.33) with the above exponential and addition. We find

$$\frac{\mathrm{d}^2 X^i_{l^*,\,m^*,\,n^*}}{\mathrm{d}t^2} = \sum_{k=1}^{3} X^k_{l^*,\,m^*,\,n^*} \sum_{\lambda,\,\mu,\,\nu} \kappa^{ik}_{\lambda,\,\mu,\,\nu} \, \mathrm{i} \mathrm{x} \mathrm{p} \left(-\frac{l^*\lambda}{L} - \frac{m^*\mu}{M} - \frac{n^*\nu}{N}\right). \qquad (3.35)$$

Observe that only three normal co-ordinates appear on the right. Indeed, if we let i run from 1 to 3 we have a closed system of three equations in three unknowns with all the complications tucked away into the nine coefficients on the right; in other words, we can

write down for each value of the wave vector an equation system
of the form

$$\frac{d^2X^{(1)}}{dt^2} = \sigma^{(11)}X^{(1)} + \sigma^{(12)}X^{(2)} + \sigma^{(13)}X^{(3)},$$

$$\frac{d^2X^{(2)}}{dt^2} = \sigma^{(21)}X^{(1)} + \sigma^{(22)}X^{(2)} + \sigma^{(23)}X^{(3)},$$

$$\frac{d^2X^{(3)}}{dt^2} = \sigma^{(31)}X^{(1)} + \sigma^{(32)}X^{(2)} + \sigma^{(33)}X^{(3)}.$$

The system yields three circular frequencies ω which can be
written as roots of a secular equation

$$\begin{vmatrix} \sigma^{(11)} + \omega^2 & \sigma^{(12}} & \sigma^{(13)} \\ \sigma^{(21)} & \sigma^{(22)} + \omega^2 & \sigma^{(23)} \\ \sigma^{(31)} & \sigma^{(32)} & \sigma^{(33)} + \omega^2 \end{vmatrix} = 0. \qquad (3.36)$$

The reason we get three frequencies is because elastic waves have
three states of polarization. Generally, the cubic equation (3.36)
must be solved to find the correct polarizations for given wave
length. However, in crystals of high symmetry, the equation will
factor for propagation in certain special directions. We shall not
discuss fully how to get the nine coefficients $\sigma^{(ik)}$; the problem is
not always as complicated as it looks, because often the coupling
constants $\kappa^{ik}_{\lambda,\mu,\nu}$ in (3.33) and (3.35) are different from zero only for
very low values of $\pm\lambda$, $\pm\mu$, $\pm\nu$ (maybe 0, 1 and 2). Hence the long
sums in (3.35) have only something like a dozen terms. The end
result of the calculation will be three frequencies and three polarizations as functions of the wave vector κ; this wave vector is implicitly
contained in (3.30). To make it apparent one must give to the large
bracket of that expression the form $-\kappa\cdot\mathbf{r}$, with $\mathbf{r}$ given by (1.2).
One finds

$$\kappa(l^*, m^*, n^*) = \frac{l^*\mathbf{a}^*}{L} + \frac{m^*\mathbf{b}^*}{M} + \frac{n^*\mathbf{c}^*}{N}, \qquad (3.37)$$

where the reciprocal lattice vectors $\mathbf{a}^*$, $\mathbf{b}^*$, $\mathbf{c}^*$ are given by (2.9)
and (2.10). For crystals which are not simple space lattices, obvious
complications will have to be added to this picture. Thus, for
instance, diamond or germanium will have six frequencies for
each wave vector because there are two atoms per primitive cell.

FREQUENCY SPECTRUM OF THE PHONONS

At present, there are few fields of physics in which the detailed structure of the vibration spectrum of a solid is of importance. There is, however, one quantity $g(f)$, the *number of frequencies per unit frequency range*, which strongly influences its thermal properties, as we shall see below. Some of the ingenuity of mathematical physicists has therefore gone into computing or guessing this quantity for actual crystals. The starting point for such a calculation is the observation, obvious from (3.11) or (3.37), that the distribution of modes in wave number space is uniform, filling the first Brillouin zone. Passage from this space to a frequency scale can then be effected by differential geometry.

We may understand the idea behind the general procedure by computing first the number $g(f)$ of frequencies per unit frequency range for the model of fig. 3.1. In this case, the variable p is a convenient one; we see from (3.15) that it ranges from 0 to $\frac{1}{2}N$ and has two modes per unit range. Combining this with the definition of $g(f)$ we get for the number of modes within dp

$$2\,dp = 2\frac{df}{df/dp} = g(f)\,df.$$

Hence $g(f) = 2/(df/dp),$

which yields with (3.8)

$$g(f) = \frac{2N}{[(k/M) - \pi^2 f^2]^{\frac{1}{2}}}. \qquad (3.38)$$

The curve is shown in fig. 3.6. It rises to infinity at the frequency maximum and has a finite value at low frequency. We could derive this latter value also from a continuum theory, using (3.17) and (3.19).

In a three-dimensional case the corresponding starting point is (3.37) which predicts a simple density of modes in a space with axes l^*, m^*, n^*; by (3.37) this is the wave number space. Each spectral branch has a separate frequency assignment which can be taken care of either by several $g(f)$'s or by a summation formula of the form

$$g(f)\,df = \sum_{i=1}^{3} \iiint_{\Omega_i(f)} dl^*\,dm^*\,dn^*. \qquad (3.39)$$

Here $\Omega_i(f)$ is the quasi-infinitesimal volume between f and $f+df$ for the branch i. The easiest way to carry out the prescription (3.39) is to transform the right-hand side so as to make f one of the variables of integration, and then to drop the integration over f.

The calculation as outlined by (3.39) is sufficiently tedious to suggest simplifying assumptions for quick results. Historically the oldest is due to Einstein who assumed that $g(f)$ may be approximated

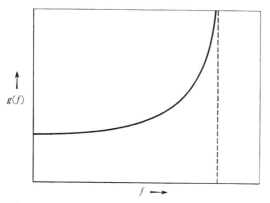

$g(f)$

$f \longrightarrow$

Fig. 3.6. Frequency distribution function versus frequency for the model of fig. 3.1. The maximum frequency is also a singularity of the distribution function.

by δ-functions at one or several positions. This was followed by the one of Debye which has proved so convenient that it is still in common use today.[†] Debye was mainly concerned with the acoustical branch of the spectrum (for the optical parts, Einstein's assumption has to be used to supplement it). His observation was that the long-wave end of it is nothing else but the longitudinal and transverse elastic vibrations of the medium, which can be understood on a continuum basis. If one assumes in addition that the medium is quasi-isotropic then the entire spectrum can be described by two velocities of propagation c_T and c_L for the transverse and the longitudinal waves. It is to be understood, however, that this is an additional assumption which is not even true for cubic crystals. But if it is made, then we get for the frequency of the longitudinal wave, whose wave vector is given by (3.37), the formula

$$f_{L,l^*,m^*,n^*} = c_L \left| \kappa_{l^*,m^*,n^*} \right|. \qquad (3.40)$$

† P. Debye, *Ann. Phys., Lpz.*, **39** (1912), 789.

We now prepare the integration (3.39) by transforming it to a volume integral in κ-space. We have from (3.37)

$$\iiint F(\kappa)\,d\kappa = \frac{\mathbf{a}^* \cdot (\mathbf{b}^* \times \mathbf{c}^*)}{LMN} \iiint F(\kappa)\,dl^*\,dm^*\,dn^*.$$

We apply to this the identity of problem 4, Chapter 2, and get

$$\iiint F(\kappa)\,dl^*\,dm^*\,dn^* = \Omega \iiint F(\kappa)\,d\kappa; \qquad (3.41)$$

here Ω is the Born–von Karman volume, and the integration on the right goes over the first Brillouin zone. For our particular case (3.39) we can transform further with (3.40)

$$\iiint\limits_{\Omega(f)} dl^*\,dm^*\,dn^* = \Omega \iiint\limits_{\Omega(f)} d\kappa$$

$$= \Omega 4\pi \int_{\Omega(f)} \kappa^2\,d\kappa$$

$$= \frac{4\pi\Omega f^2}{c_L^3},$$

and hence
$$g_L(f) = \frac{4\pi\Omega f^2}{c_L^3}. \qquad (3.42\,a)$$

It is seen that the result is a function measuring the number of frequencies per unit frequency range *and unit volume*. This has nothing to do with the Debye model because it is already a consequence of the general identity (3.41). Thus only the proportionality to f^2 in equation (3.42 a) is characteristic of the model. The formula for the transverse $g_T(f)$ runs the same way and yields

$$g_T(f) = \frac{8\pi\Omega f^2}{c_T^3} \qquad (3.42\,b)$$

because of the two states of polarization. The expressions (3.41) and (3.42) have to be supplemented by the prescription that the $g(f)$'s drop to zero at the edge of the first Brillouin zone. In applications, it is sometimes convenient to replace the actual zone by a sphere of equal volume in κ-space; the error introduced thereby is probably not worse than those already committed. The maximum frequency F_L is then obtainable from the relation

$$\int_0^{F_L} g_L(f)\,df = LMN,$$

which, when substituted into (3.42a) yields

$$F_L = c_L \left(\frac{3}{4\pi\omega}\right)^{\frac{1}{3}}, \qquad (3.43)$$

where ω is the volume of the primitive cell. The curve which results from (3.41) and (3.43) is shown in fig. 3.7.

The problem of computing $g(f)$ for actual crystals is of such mportance that efforts have been made to go beyond the Debye

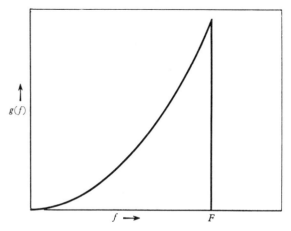

Fig. 3.7. Debye model of the frequency distribution function of a solid: a parabola cut off at a frequency maximum F.

model. The early approach was to compute numerically the roots of (3.36) for a sufficient number of wave vectors so that a good approximation to $g(f)$ could be had by direct count. Such methods yield, of course, only a vague outline of the spectrum, and it is thus not too surprising that the early curves were rounded and rather dull-looking affairs. However, we shall now outline the topological considerations which show that this is far from being the case for the actual spectrum.

First, there are well-known features at the positions of frequency maxima and minima. Equations (3.35), (3.36) and (3.37) show that we get f^2 as periodic function of the wave vector $\kappa(l^*, m^*, n^*)$. Near a maximum or a minimum of ω, we can impose upon the three components of κ a linear transformation so as to end up with three Cartesian co-ordinates x, y, z, which vanish for the extremum and for which

$$f^2 = f_0^2 \pm (x^2 + y^2 + z^2). \tag{3.44}$$

We can measure the volume (3.39) in terms of $dx\,dy\,dz$ because it differs from the original one only by a non-vanishing real factor. Similar transformations will be allowed subsequently on x, y, z in order to simplify notation. There are then three types of situations which may arise from (3.44):

(a) *The bottom of the acoustical branch.* We have $f_0 = 0$, and hence

$$f = (x^2 + y^2 + z^2)^{\frac{1}{2}}. \tag{3.45}$$

To follow the recipe subsequent to (3.39) we transform

$$x = f \sin \vartheta \cos \phi,$$
$$y = f \sin \vartheta \sin \phi,$$
$$z = f \cos \vartheta,$$

and get

$$g(f) \propto f^2. \tag{3.46}$$

This feature is incorporated in the Debye model as seen in (3.42). It makes the model basically correct for long waves, and shows that the inaccuracies of detail in (3.42) can be eliminated at the low frequency end of the spectrum.

(b) *Other minima.* We have

$$f = f_0 + x^2 + y^2 + z^2. \tag{3.47}$$

The transformation is

$$x = (f - f_0)^{\frac{1}{2}} \sin \vartheta \cos \phi,$$
$$y = (f - f_0)^{\frac{1}{2}} \sin \vartheta \sin \phi,$$
$$z = (f - f_0)^{\frac{1}{2}} \cos \vartheta,$$

and the result

$$g(f) \propto (f - f_0)^{\frac{1}{2}}. \tag{3.48}$$

This is a rise from zero with vertical slope, as shown in fig. 3.8. The rise can occur on a normal background in which case it may take the appearance of fig. 3.9.

(c) *Any maximum.* In this case, (3.44) becomes

$$f = f_0 - (x^2 + y^2 + z^2). \tag{3.49}$$

The transformation is the same as under (b) except for a reversal of sign. We get

$$g(f) \propto (f_0 - f)^{\frac{1}{2}}. \tag{3.50}$$

This is exhibited in fig. 3.10. It can again occur on a continuous background.

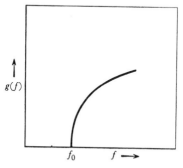

Fig. 3.8. Structural aspect of the frequency distribution function at a frequency minimum.

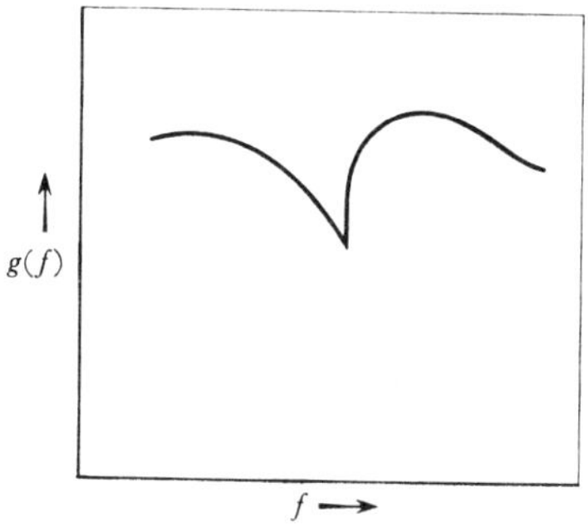

Fig. 3.9. Same as preceding figure, on a continuous background.

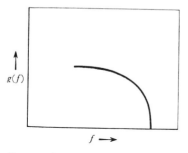

Fig. 3.10. Structural aspect of the frequency distribution function at a frequency maximum.

In addition to maxima and minima, there may be saddle points in the interior of frequency bands. In the neighbourhood of such a point the surfaces of constant frequency pass from being one-sheeted to two-sheeted hyperboloids; at the critical frequency itself, the surface is a cone (fig. 3.11). If the sequence shown is for rising frequency then we may write, as in (3.44)

$$f = f_0 + z^2 - x^2 - y^2. \tag{3.51}$$

We introduce the transformation

below f_0:

$$x = (f_0-f)^{\frac{1}{2}} \cosh \xi \cos \phi,$$

$$y = (f_0-f)^{\frac{1}{2}} \cosh \xi \sin \phi,$$

$$z = (f_0-f)^{\frac{1}{2}} \sin \xi.$$

above f_0:

$$x = (f-f_0)^{\frac{1}{2}} \sinh \zeta \cos \phi,$$

$$y = (f-f_0)^{\frac{1}{2}} \sinh \zeta \sin \phi,$$

$$z = (f-f_0)^{\frac{1}{2}} \cosh \zeta.$$

$$dx\,dy\,dz =$$

$$\tfrac{1}{2}(f_0-f)^{\frac{1}{2}} \cosh \xi\, df\, d\xi\, d\phi. \qquad \tfrac{1}{2}(f-f_0)^{\frac{1}{2}} \sinh \zeta\, df\, d\zeta\, d\phi.$$

Fig. 3.11. Picture of the surfaces of constant frequency in wave number space in the neighbourhood of a saddle point.

Care must be taken here to limit the integration to some fixed distance from the saddle point. The most convenient form for this restriction is

$$x^2+y^2+z^2 < R^2,$$

which yields

$$\cosh 2\xi < \frac{R^2}{f_0-f},$$

$$0 < \sinh \xi < \left[\frac{1}{2}\left(\frac{R^2}{f_0-f}-1\right)\right]^{\frac{1}{2}}.$$

$$\cosh 2\zeta < \frac{R^2}{f-f_0},$$

$$1 < \cosh \zeta < \left[\frac{1}{2}\left(\frac{R^2}{f-f_0}+1\right)\right]^{\frac{1}{2}}.$$

We now get $g(f)$ by our usual prescription of dropping df in the element of volume and integrating the remainder. The integral over $d\phi$ yields 2π and the other angular integral is worked out in the last equations above. We find

below f_0:

$$g(f) \propto 2\pi \{\tfrac{1}{2}(R^2+f-f_0)\}^{\frac{1}{2}}. \tag{3.52a}$$

above f_0:

$$g(f) \propto 2\pi [\{\tfrac{1}{2}(R^2+f-f_0)\}^{\frac{1}{2}} - (f-f_0)^{\frac{1}{2}}]. \tag{3.52b}$$

The first term is common to the two formulas; it is finite and continuous with all its derivatives at $f = f_0$. Hence it simply yields a continuous background. On the other hand, the second term in (3.52b), while zero itself at $f = f_0$, has an infinite derivative and

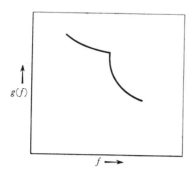

Fig. 3.12. Structural aspect of the frequency distribution function at a saddle point having the form of equation (3.51).

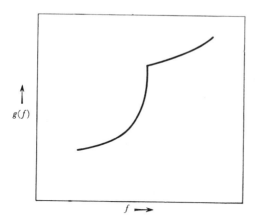

Fig. 3.13. Structural aspect of the frequency distribution function at a saddle point having the form of equation (3.53).

thereafter falls away with large negative slope. The resulting behaviour is shown in fig. 3.12. In the same way, if the sequence of fig. 3.11 is reversed, the picture of fig. 3.13 results.

Van Hove† first realized the nature of the influence of saddle points on spectral distribution. He also proved at the same time that

† L. Van Hove, *Phys. Rev.* **89** (1953), 1189.

at least one of either type of saddle point occurs in an actual frequency band. The argument is topological and in its essence easy to see. f is periodic in κ-space and therefore has at least one maximum and one minimum in each primitive cell of the reciprocal lattice. Let two such neighbouring maxima be A and B (fig. 3.14). Now stack a set of parallel planes between A and B including the ends. On each plane there will be a set of maxima too (make the planes simple lattice planes to guarantee periodicity). A maximum on the planes containing A or B will be A or B. In between, other

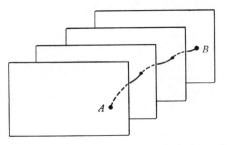

Fig. 3.14. Illustration of Van Hove's topological proof for the presence of saddle points in actual bands.

maxima will arise which, by continuity, will form a curve connecting A and B. Somewhere along the curve the maximum will be smallest. At this point all first derivatives vanish and

$$f = f_0 + x^2 + y^2 - z^2. \tag{3.53}$$

This is a saddle point yielding a $g(f)$ as shown in fig. 3.13. Similarly, an argument on minima will yield a saddle point of the type (3.51) and hence a $g(f)$ behaving as in fig. 3.12. In order to understand this argument fully, the reader is invited to prove for himself, that for a periodic landscape of hills and valleys in two dimensions, there must be at least one saddle point in each elementary parallelogram. In this case, his direct intuition will come to his aid.

If we gather all the information together we find that the simplest possible vibration spectrum has the form shown in fig. 3.15. It is doubtful whether it occurs for any substance. Already for the face-centred cubic structure, for which the first Brillouin zone is the truncated octahedron shown in fig. 1.31, there are by symmetry more than four positions at which all derivatives of f^2 with respect to κ must vanish. There are two different kinds of face centres on the polyhedron, each of which yields a saddle point of the type (3.53), and there are two kinds of edges, each of which has in the middle a saddle point of the type (3.51). If we make the natural

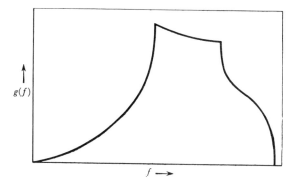

Fig. 3.15. Simplest possible aspect of the frequency distribution function of a solid.

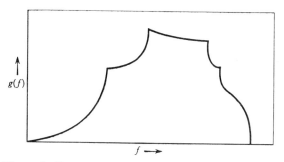

Fig. 3.16. Frequency distribution function for one polarization in the face-centred cubic structure.

assumption that f goes up with $|\kappa|$, a $g(f)$ as shown in fig. 3.16 will result. To represent the actual $g(f)$ for all three polarizations three curves of this type must be superimposed. Some of the singular points will coincide by symmetry, but even so the resultant $g(f)$ will have a good deal of structure. Further complications will arise if we deal with a more complicated crystal structure than the one discussed.

THE SPECIFIC HEAT OF SOLIDS

The law of Dulong–Petit which ascribes to all monatomic solids a molar specific heat $3R$ is explained in terms of a classical theory of lattice vibrations. For there are $3A$ normal modes in one mole (A being Avogadro's number), and each one has classically the energy kT; this yields the desired answer.

However, the law of Dulong–Petit is only a limiting law for elevated temperature. As the temperature of a solid is lowered its heat capacity drops and becomes zero at the absolute zero of temperature. In a qualitative way this drop is predictable from Nernst's heat theorem, which is also called the third law of thermodynamics.† The theorem asserts that the entropy of any pure chemical substance is zero at absolute zero. The theorem is partly nullified by the sluggishness of transformations at low temperatures; there is, however, a weak form of it, which is rigorously true, and which is sufficient for the present discussion, namely to assert that the entropy of any substance is finite at absolute zero. Since entropy is defined as $\int \dfrac{C(T)}{T} \, dT$, where C is a heat capacity, a finite value for this integral entails the consequence that $C(T)$ is zero when T is zero. Nernst's heat theorem could never be implemented by atomic models in classical physics. The vanishing of $C(0)$ is a quantum effect, and occurs quite generally for any degree of freedom. As soon as the temperature drops sufficiently, so that kT becomes less than the spacing of the quantized levels, the specific heat of this degree of freedom begins to depart from equipartition and finally tends to zero. Degrees of freedom of this kind are called *dormant*; the best known case are the dormant rotational degrees of freedom of monatomic and diatomic gaseous molecules.

In the case of solids the degrees of freedom we are talking about are the lattice vibrations discussed earlier in this chapter. We have seen that they are equivalent to uncoupled harmonic oscillators if their amplitudes are not too large. A correct theory of the heat capacity of solids must take into account that these oscillators are quantized. The levels are given by

$$E_n = (n + \tfrac{1}{2}) hf, \qquad (3.54a)$$

† W. Nernst, *Göttinger Nachr.* **1** (1906), 1.

with $\qquad\qquad n = 0, 1, 2, 3 \dots$ $\qquad\qquad$ (3.54b)

From Boltzmann's law of energy distribution, the probability P_n of finding a mode in an energy state E_n is

$$P_n = \frac{e^{-E_n/kT}}{\sum\limits_{\nu=0}^{\infty} e^{-E_\nu/kT}}.$$ (3.55)

Therefore the mean energy $\langle E \rangle$ in the mode becomes

$$\langle E \rangle = \sum_\nu P_\nu E_\nu = \frac{\sum\limits_{\nu=0}^{\infty} (\nu + \tfrac{1}{2}) hf\, e^{-\nu hf/kT}}{\sum\limits_{\nu=0}^{\infty} e^{-\nu hf/kT}}$$

$$= \tfrac{1}{2}hf + kT^2 \frac{d}{dT}\left[\ln\left(\sum\limits_{\nu=0}^{\infty} e^{-\nu hf/kT}\right)\right]$$

$$= \tfrac{1}{2}hf - kT^2 \frac{d}{dT}[\ln(1 - e^{-hf/kT})],$$

and hence finally

$$\langle E \rangle = \tfrac{1}{2}hf \coth\frac{hf}{2kT}.$$ (3.56)

The appearance of this energy versus temperature curve is shown in fig. 3.17. The contribution to the specific heat is the slope of this curve; it rises from zero at absolute zero to the constant demanded by the law of Dulong and Petit, as shown in fig. 3.18. This curve, first proposed by Einstein[†] for the specific heat of solids, is in qualitative agreement with the facts. However, the drop-off at low temperature is too steep.

A more careful formulation of the statistical problem for solids brings in the quantity $g(f)$ discussed earlier in this chapter. The normal modes of vibration of a solid were shown to be a set of uncoupled harmonic oscillators having a frequency distribution $g(f)$. From this result and (3.56) we get for the total energy U of the solid

$$U = \tfrac{1}{2}h \int_0^F fg(f) \coth\frac{hf}{2kT}\, df.$$ (3.57)

From the point of view of this equation, Einstein's formula is an

† A. Einstein, *Ann. Phys., Lpz.,* **22** (1906), 180, 800.

approximation to the truth, in which $g(f)$ is replaced by a δ-function at a fixed frequency. The Debye model takes for $g(f)$ the sum of (3.41) and (3.42). Previous discussion has shown these expressions

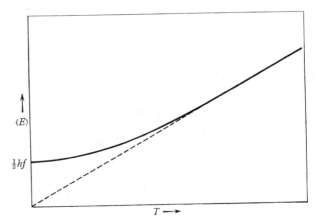

Fig. 3.17. Average energy versus temperature of a quantized harmonic oscillator: Einstein model of a solid.

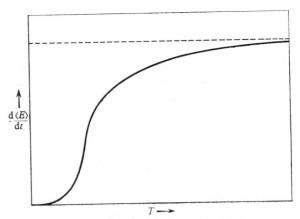

Fig. 3.18. Specific heat versus temperature for the Einstein model of a solid.

to be exact at the low frequency end of the acoustical spectrum (equation (3.46)). At the same time, the Debye theory counts the number of normal modes correctly for an atomic solid crystallizing in a simple lattice; it reaches thereby the correct high temperature limiting law of Dulong and Petit. In this way, the Debye theory comes out to be an excellent all-round approximation for simple solids. In order to avoid needless complication we shall simplify the theory still further, neglecting even the distinction between longitudinal and transverse waves. We then get from (3.41), (3.42) and (3.43)

$$g(f) = \frac{9Nf^2}{F^3}.$$ (3.58)

Inserting this into (3.57) we find for the energy U

$$U = \frac{9Nh}{2F^3} \int_0^F f^3 \coth \frac{hf}{2kT} \, df.$$

The appearance of this formula is usually simplified by introduction of the so-called *Debye temperature* Θ:

$$hF = k\Theta.$$ (3.59)

U then takes the form

$$U = \frac{9NkT^4}{2\Theta^3} \int_0^{\Theta/T} x^3 \coth \tfrac{1}{2}x \, dx,$$ (3.60)

or, splitting off the zero point energy,

$$U = \tfrac{9}{8}Nk\Theta + \frac{9NkT^4}{\Theta^3} \int_0^{\Theta/T} \frac{x^3 \, dx}{e^x - 1}.$$ (3.61)

The equations show that U equals its classical value $3NkT$, multiplied with a universal function of T/Θ only. This means that its temperature derivative, the molar specific heat, is a universal function of T/Θ. This prediction is brilliantly verified for a large number of monatomic solid substances. Even though these substances differ greatly from each other in their molar specific heats, it comes out that these values can be made to coincide with the universal Debye curve by one simple scaling operation on the temperature. In fig. 3.19 this curve is shown, together with experimental points for two extreme members, lead and diamond. The Debye temperatures of these two materials are 88° and 1850° respectively, yet their specific heat curves are seen to follow the same relationship.

The universal function (3.61), which is exhibited in fig. 3.19 through its derivative, is not expressible in closed form in terms of known functions, and thus has to be handled through numerical tables. The law does, however, contain simple limiting laws which are worth considering in greater detail. These laws hold respectively in the temperature regions in which T is very much less and very much greater than Θ.

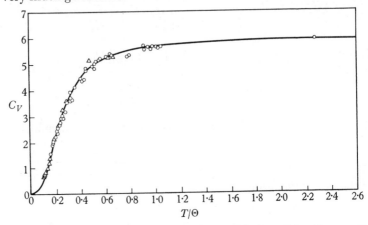

Fig. 3.19. Specific heat curve for the Debye model of a solid, with measured points for lead ($\bigcirc$) and diamond ($\triangle$). The abscissa scale contains the Debye temperature Θ and is adjusted to bring the experimental points as well as possible on the theoretical curve (after Schrödinger).

To derive the low temperature formula we observe that if

$$T \ll \Theta, \qquad (3.62\,a)$$

the upper limit in the integral in (3.61) is large and the integrand for values beyond that limit small. We may therefore replace the limit by $+\infty$. The integral then becomes

$$\int_0^\infty x^3\,\mathrm{d}x(\mathrm{e}^{-x}+\mathrm{e}^{-2x}+\mathrm{e}^{-3x}+\ldots) = 6\left[1+\frac{1}{2^4}+\frac{1}{3^4}+\ldots\right] = \frac{\pi^4}{15},$$

and hence

$$U = \tfrac{9}{8}Nk\Theta + \frac{3\pi^4 NkT^4}{5\Theta^3}. \qquad (3.62\,b)$$

This yields for the specific heat per mole

$$C = \frac{12\pi^4}{5}R\left(\frac{T}{\Theta}\right)^3. \qquad (3.63)$$

This is the so-called Debye T^3-law. It is a rigorously correct law for a suitably chosen Θ, because the Debye $g(f)$ is correct for low frequencies, as shown by equation (3.46). In fact, this equation shows that Θ can be given an exact meaning. Namely, Θ is the number which, when substituted into (3.59) and from there into (3.58), yields the correct value of the multiplier in (3.46). This is purely a problem of acoustics because the formula (3.46) deals with very long waves only. According to acoustical reasoning, if Θ is proportional to F, it is also proportional to an acoustical velocity, and that, in turn, is proportional to the square root of an elastic modulus. However, to find the proper modulus one has to take the correct average over all directions of propagation and the three polarizations; this requires the notions of elasticity theory for anisotropic media which will not be discussed here. The resultant Θ so calculated from the elastic constants should be in agreement with the Θ obtained from (3.63). This is true in general, but occasional discrepancies are known. They must arise from a misinterpretation of specific heat data.

To derive a specific heat law valid when

$$T \gg \Theta \qquad (3.64a)$$

we start from (3.60) and observe that, now, the variable x remains small. The coth can therefore be expanded:

$$\coth \tfrac{1}{2}x = \frac{2}{x} + \tfrac{1}{6}x + O(x^3),$$

yielding
$$U = 3NkT + \frac{3}{20}\frac{Nk\Theta^2}{T} + O\left(\frac{Nk\Theta^4}{T^3}\right), \qquad (3.64b)$$

and
$$C = 3R - \tfrac{3}{20}R\left(\frac{\Theta}{T}\right)^2 + \dots \qquad (3.65)$$

These formulas have not the same standing as (3.63) because the approximation neglects essential features of the frequency spectrum. A simple way to check on the applicability of the Debye theory to a given substance is to compute an empirical Θ for low and high temperature from (3.63) and (3.65) respectively, and to see how they compare. The same thing can be extended to intermediate temperatures, and thus yields an empirical curve for Θ versus T. The result is, that while Θ is not a constant, it varies remarkably little, often by not more than 20 per cent.

CONSERVATION LAW FOR CRYSTAL MOMENTUM

Next in importance to the quantity $g(f)$ are the conservation laws to which lattice vibrations give rise when they interact with other waves. These laws play an important role in the scattering of electromagnetic or matter waves by the lattice waves excited in a thermally agitated solid. The conservation laws are not a peculiarly quantum mechanical phenomenon. However, the quantum reasoning is more straightforward and different from the classical approach. In view of this, it is worthwhile to discuss the basic phenomenon in both languages.

In the classical approach the conservation laws arise if one considers the modulation produced on a scattering phenomenon of the type (2.2) if the scattering centres are themselves in motion according to a mode of vibration. Let us assume therefore that the atoms are displaced according to one particular mode of vibration, with the polarization chosen properly according to (3.36). The displacement ξ, as explained earlier, may then be obtained as function of position by taking one X, letting it vary in time according to (3.10), and taking the real part of the resultant x obtained from (3.30). This yields

$$\xi(\mathbf{x}) = \mathbf{e}\, a \cos 2\pi(\kappa \cdot \mathbf{x} - ft), \tag{3.66}$$

where $\mathbf{e}$ is a unit polarization vector and a an amplitude factor. The actual position $\mathbf{r}'$ of the atom located normally at $\mathbf{r}$ is thus

$$\mathbf{r}' = \mathbf{r} + \xi(\mathbf{r}). \tag{3.67}$$

The problem is to calculate the coherent scattering of a wave by atoms displaced according to (3.67). For this purpose let us assume an incoming wave of the form

$$A = A_0\, e^{-2\pi i(\mathbf{k}_0 \cdot \mathbf{x} - g_0 t)}, \tag{3.68}$$

which at the instantaneous atomic position $\mathbf{x} = \mathbf{r}'$ becomes

$$A(\mathbf{r}) = A_0 \operatorname{ixp}[g_0 t - \mathbf{k}_0 \cdot (\mathbf{r} + \xi(\mathbf{r}))],$$

where the abbreviation (3.12) has been employed. For $\mathbf{k}_0 \cdot \xi$ small, this is equal to

$$\begin{aligned}
A(\mathbf{r}) &= A_0\, e^{2\pi i(g_0 t - \mathbf{k}_0 \cdot \mathbf{r})}[1 - i\,\mathbf{k}_0 \cdot \xi(\mathbf{r})] \\
&= A_0 \operatorname{ixp}(g_0 t - \mathbf{k}_0 \cdot \mathbf{r}) \\
&\quad - \tfrac{1}{2} i A_0\, \mathbf{k}_0 \cdot \mathbf{e}\, a \operatorname{ixp}[(g_0 - f)\, t - (\mathbf{k}_0 - \kappa) \cdot \mathbf{r}] \\
&\quad - \tfrac{1}{2} i A_0\, \mathbf{k}_0 \cdot \mathbf{e}\, a \operatorname{ixp}[(g_0 + f)\, t - (\mathbf{k}_0 + \kappa) \cdot \mathbf{r}].
\end{aligned}$$

This is the expected modulation of the received signal by the moving receiver: in addition to the original signal we have two sidebands for which

$$g_1 = g_0 - f, \quad \mathbf{k}_1 = \mathbf{k}_0 - \boldsymbol{\kappa},$$

and

$$g_2 = g_0 + f, \quad \mathbf{k}_2 = \mathbf{k}_0 + \boldsymbol{\kappa}.$$

The amplitudes of these sidebands are proportional to the received signal and to the amplitude a of the excited lattice vibration. The condition for coherent scattering by all atoms now proceeds as in Chapter 2 and ends up with the generalization of (2.12), namely either

$$g = g_0 - f, \tag{3.69}$$

$$\mathbf{k} = \mathbf{k}_0 - \boldsymbol{\kappa} + \mathbf{r}^*_{LMN}, \tag{3.70}$$

or

$$g = g_0 + f, \tag{3.71}$$

$$\mathbf{k} = \mathbf{k}_0 + \boldsymbol{\kappa} + \mathbf{r}^*_{LMN}. \tag{3.72}$$

In the quantum derivation of the same equations the basic physical fact is the coupling of the scattered wave field to the phonon field, and the process appears as scattering with simultaneous emission or absorption of a phonon. As a typical example of the type of coupling to be expected we may take the deformation potential theory of Bardeen and Shockley.[†] This theory discusses a potential energy shift U for the scattered particle proportional to the volume strain:

$$U = u \nabla \cdot \boldsymbol{\xi}, \tag{3.73}$$

where u is a constant. To analyse elementary processes we take again $\boldsymbol{\xi}$ in the form (3.66). This yields then

$$U = 2\pi u a \, \mathbf{e} \cdot \boldsymbol{\kappa} \sin 2\pi (\boldsymbol{\kappa} \cdot \mathbf{x} - ft). \tag{3.74}$$

In the simplest approximation, scattering is controlled by the matrix element of U connecting the incoming plane wave (3.68) to a specified outgoing wave of wave vector $\mathbf{k}$ and frequency g:

$$M = 2\pi A_0 a u \, \mathbf{e} \cdot \boldsymbol{\kappa} \int_\Omega e^{2\pi i (\mathbf{k} \cdot \mathbf{x} - gt)} e^{-2\pi i (\mathbf{k}_0 \cdot \mathbf{x} - g_0 t)} \sin 2\pi (\boldsymbol{\kappa} \cdot \mathbf{x} - ft) \, d\tau.$$

$$\tag{3.75}$$

† J. Bardeen and W. Shockley, *Phys. Rev.* **80** (1950), 72. W. Shockley, *Electrons and holes in semiconductors* (New York, Van Nostrand, 1955), pp. 520–31.

The two exponentials in the sine-function must be treated separately in the discussion. Thereafter, we get the equation (3.69) or (3.71) by annulling the frequency factor multiplying t; this means conservation of energy in the scattering process. Then in carrying out the spatial integral we get the condition (3.70) or (3.72). The indeterminacy arising from $\mathbf{r}^*_{LMN}$ enters because the periodic structure of the medium modifies all three waves, and the elastic wave in particular, as explained in the early part of this chapter. In other words, the expression (3.66) should in reality be multiplied with a factor having the periodicity of the lattice; such a factor may be written as a Fourier series of the form

$$\sum_{\mathbf{r}^*} u(\mathbf{r}^*)\, e^{2\pi i \mathbf{r}^* \cdot \mathbf{x}},$$

where $\mathbf{r}^*$ is the general reciprocal lattice vector (2.11). This factor, when transferred into (3.75), produces the indeterminacy indicated. Processes, in which the wave vector shifts by some $\mathbf{r}^*$ are sometimes called 'Umklapp' processes in the literature. It is reasonable to suppose them rare when all wave-lengths of the process are long. Just as in the preceding classical calculation we observe in (3.75) that the result is proportional to the product $A_0 a$; in addition, however, the present formulation yields information about the coupling constant between the two wave fields, which here equals $u\mathbf{e}\cdot\boldsymbol{\kappa}$.

It is observed that the conservation equations (3.69)–(3.72) are the same in number as the original Laue equations and therefore, for fixed $\boldsymbol{\kappa}$ and f, are solvable only under the same specialized conditions. However, the values of $\boldsymbol{\kappa}$ and f are picked out of a band-type continuum, and therefore, the extra condition is satisfied here by imposing a selection within the band.†

Application of the relations (3.69)–(3.72) is either to matter or to electromagnetic waves. If the material particles can be treated as free, $\mathbf{k}$ and g are related to the momentum $\mathbf{p}$ and the energy E through the relations

$$\mathbf{k} = \frac{\mathbf{p}}{h}, \tag{3.76}$$

$$g = \frac{E}{h}. \tag{3.77}$$

† For a picture of this selection in a particular case, see the preceding reference to W. Shockley, p. 530.

Clearly, the relations (3.69) and (3.71) then appear as energy conservation relations, with the phonon energies given by (3.23). By analogy, the relations (3.70) and (3.72) are often referred to as *momentum relations*. This cannot be entirely accurate because the presence of the vector $\mathbf{r}^*_{LMN}$ would then allow exceptions to the law of conservation of momentum. Momentum conservation is a law which can be proved in classical or quantum mechanics for any system and thus must exist over and above the relations (3.70) and (3.72); in order to apply them, however, we must include the crystal in the interacting system. The equations in question, on the other hand, refer to the motion of waves in a stratified medium. This becomes particularly clear if we set for a moment $\kappa = f = 0$, which returns us to the subject matter of Chapter 2. (3.70) and (3.72) are now identical with (2.12), one of the forms of the Bragg law. According to that law, a wave may be reflected from a lattice plane, which means that its momentum is certainly not conserved. The situation is clarified in this respect if the matter wave is tightly coupled to the lattice. As we shall see later, the equation (3.76) then no longer holds, but only the weaker condition

$$e^{2\pi i \mathbf{k} \cdot \mathbf{r}} = e^{2\pi i \mathbf{p} \cdot \mathbf{r}/h} \tag{3.78}$$

for any $\mathbf{r}$ satisfying (1.2). This equation is compatible with an expectation value of $\mathbf{p}/h$ entirely different from the value of $\mathbf{k}$. We must therefore distinguish *two* momentum conservation laws: the usual one, which must allow for crystal recoil and the law of conservation of *crystal momentum*, which holds only up to certain additive constants, and which considers the crystal a medium of propagation having periodic structure.

Application of these general principles to particular cases is beyond the scope of this book. It may only be mentioned that the waves to be scattered are generally such that it is difficult to observe simultaneously the energy and momentum shift, that is the κ and f of the elastic wave involved. If κ and $\mathbf{k}$ are comparable, f is usually too small in comparison to g, and if f and g are made comparable then only elastic waves of small κ can participate in the scattering process. An exception arises for neutrons of thermal energy at room temperature, for which both quantities are simultaneously comparable. Since it is still impossible experimentally to manipulate the bulk of the crystal vibration spectrum directly, this exception has experimental importance.†

PROBLEMS

1. Plot on the same graph two waves given by

$$y = \sin \frac{5\pi x}{6}$$

and

$$y = -\sin \frac{7\pi x}{6}$$

letting x run from 0 to 12. Observe the values of y for integer x in the two cases.

2. Calculate frequencies and wave characteristics for a problem analogous to (3.24), in which all masses are equal, but the springs are alternately weak and strong. Prove that one can again distinguish an acoustical and an optical branch with phase shifts of 0 and π respectively, provided phases are measured at the symmetry centres of the structure rather than at the mass points.

3. Suppose we have a landscape of hills and valleys in the plane which has two non-parallel periods. Show that the landscape has at least one saddle point per primitive cell.

4. Calculate for a two-dimensional crystal how $g(f)$ varies near a saddle point of a $f(\kappa_x, \kappa_y)$ plot.

5. Find the way in which an atomic p level will split up when placed in a position of (a) cubic and (b) axial symmetry.

6. Find the way in which an atomic d level will split up when placed in a position of (a) cubic, and (b) trigonal symmetry. Does your result agree with the fact that a cube has trigonal symmetry about its diagonal? If not, why not?

7. Find all possible representations of the crystal translation group in the real number field.

8. Find all possible representations in the complex number field of the symmetry group of fig. 1.47. Exclude symmetries arising from the use of the third dimension.‡

† B. N. Brockhouse and A. T. Stewart, *Rev. Mod. Phys.* **30** (1958), 236 and 250. R. S. Carter, H. Palevsky and D. J. Hughes, *Phys. Rev.* **106** (1957), 1168. B. N. Brockhouse and P. K. Iyengar. *Phys. Rev.* **111** (1958), 747.

‡ The problems in representation theory given here either involve only a module of small dimension, or else can be handled by the following consideration: the total symmetry group contains the crystal translation group as a subgroup; hence the final reduced representation modules also represent the subgroup. We can make the representation of the latter diagonal by writing the module as a number of co-ordinates (3.31), leading to diagonal matrices with elements of the type (3.32). Application of additional symmetry elements will mix certain ones of these co-ordinates and yield a first approach to the true representations. Before this answer can be accepted these representations must be checked for reducibility by individual methods.

9. Find all possible representations of the groups in problem 13, Chapter 1.

10. Carry out the same analysis for the group of problem 14, Chapter 1.

BIBLIOGRAPHY

WAVE PHENOMENA IN SOLIDS

L. BRILLOUIN. *Wave Propagation in Periodic Structures.* (New York, McGraw-Hill, 1946.)

LATTICE DYNAMICS

J. DELAUNAY. The theory of specific heats and lattice vibrations. *Solid State Physics*, 2 (1956), 219–303.

M. BORN and M. GÖPPERT-MAYER. Dynamische Gittertheorie der Kristalle. *Handb. d. Phys.* XXIV, Part 2 (Berlin, Springer, 1933), pp. 623–794.

M. BORN and K. HUANG. *Dynamical Theory of Crystal Lattices.* (Oxford, Clarendon Press, 1954.)

A. A. MARADUDIN, P. MAZUR, E. W. MONTROLL and G. H. WEISS. Remarks on the vibrations of diatomic lattices. *Rev. Mod. Phys.* 30 (1958), 175.

REPRESENTATION OF GROUPS

H. EYRING, J. E. WALTER and G. E. KIMBALL. *Quantum Chemistry*, chapter 10. (New York, Wiley, 1944.)

MORTON HAMERMESH. *Group Theory and its Application to Physical Problems.* (New York, Addison-Wesley, 1961.)

J. S. LOMONT. *Applications of Finite Groups.* (New York, Academic Press, 1959.)

CHAPTER 4

CO-OPERATIVE PHENOMENA IN SOLIDS

It is a general tendency of physical theory to reduce complex problems to the consideration of a large number of independent entities, each with a small number of degrees of freedom. The reduction of the general dynamics of solids to the motion of independent phonons is of this type. The previous chapter has shown that a fair amount of empirical information can be accounted for in this way. However, some of the most characteristic features of solids do not fit into this approach; among them is the co-operative action of many particles to create an ordered structure. This aspect of physics is closer to biology than any of its other branches, and one may hope to learn from its study something about organized structures in general.

Changes of phase are among the most easily observed forms of co-operative action. This becomes particularly clear if we think of these changes as occurring in the direction of decreasing temperature. A relatively chaotic form of a substance, perhaps a gas, is then seen to lose energy until the mutual attractive forces of the molecules can make themselves felt. A succession of structures is evolved thereby in which the energy of the assembly is lowered in steps. With each step a higher degree of organization, which means a state of lower entropy, is simultaneously achieved. Most of these changes take place abruptly, at a fixed temperature T_0 for a given pressure. This temperature and the latent heat H_0 are easily accessible to observation. They are related to the change in entropy ΔS and the change in energy (more exactly enthalpy) ΔU through the thermodynamic formulas

$$\Delta U = H_0, \qquad (4.1\,a)$$

$$\Delta S = H_0/T_0. \qquad (4.1\,b)$$

A theoretical estimate of these quantities is difficult for liquids; they represent an intermediate state of organization which is hard to grasp; but for gases and solids such an estimate is often feasible.

Comparison of such estimates with actual measurements can then lead in turn to a modification of an oversimplified picture. For instance, it is found that ice retains a certain amount of entropy to very low temperatures. This is explained by assuming that while ice has a quartz-like structure (see fig. 1.42) with the oxygen atoms playing the role of the silicon atoms in quartz, the hydrogens do not bond the oxygens into the rigid kind of linkage that the oxygens of silica give to the latter; instead, they have two alternate positions available, one close to one of the oxygens, one close to the other. This produces a multiplicity of arrangements of low energy. The number W of these arrangements can be calculated from straight statistics. It turns out to be, approximately†

$$W = (\tfrac{3}{2})^N,$$

where N is the number of water molecules in the crystal. This yields with

$$S = k \ln W \tag{4.2}$$

a low temperature entropy S equal to

$$S = R \ln (\tfrac{3}{2}) = R \times 0.4055. \tag{4.3}$$

For most structural transformations of solids it is difficult to go beyond this static kind of information because the very abruptness of the change prevents us from observing the interacting forces at work. It is true that there is sometimes information available on nucleation, growth rates, growth habits, etc., but this type of information is not analysed easily by the theory because of the difficulty of interpreting rate processes in general. Theoretical interest has therefore shifted to transitions in which the solid evolves from one state of order to another through an uninterrupted range of intermediates, while remaining at all stages in thermodynamic equilibrium.

PARAMAGNETISM IN NON-METALLIC SOLIDS

As a preliminary to the study of co-operative ordering, it should be emphasized that ordering will often take place in a solid without

† L. Pauling, *J. Amer. Chem. Soc.* **57** (1935), 2680.

co-operative action. The dying down of lattice vibrations at low temperature which was discussed in the preceding chapter could be considered such a case. Let us take up here a more clear-cut example of this, the magnetization of paramagnetic materials at low temperature.

Many paramagnetic salts are close approximations to a set of localized spins, each of which responds independently to an applied magnetic field. If nature does not do the uncoupling well enough we can go to artificially made salts, which contain the paramagnetic ion in high dilution in a non-magnetic matrix (for example, chromium ions substituting for a small fraction of the aluminium in alum). Any such paramagnetic ion is then a magnetic dipole whose moment along any direction is quantized by the quantization rules of angular momentum. Let us call the value of the moment μ and the angular momentum quantum number j. The interaction energy E of the ion with an applied magnetic field is then

$$E = -\mu H(m/j), \tag{4.4a}$$

where
$$m = -j, \ -j+1, \ -j+2, \ ..., \ j-1, \ j. \tag{4.4b}$$

m is the magnetic quantum number of the ion. The mean moment $\langle\mu\rangle$ along the field, which is connected with the mean energy $\langle E\rangle$ through

$$\langle E\rangle = \langle\mu\rangle H, \tag{4.5}$$

is then a result of a statistical average over the magnetic quantum number m:

$$\langle\mu\rangle = \mu \sum_{m=-j}^{+j} P(m)\frac{m}{j}. \tag{4.6a}$$

By Boltzmann's theorem the probabilities equal

$$P(m) = \frac{\exp{(m\mu H/jkT)}}{f_j(\mu H/kT)}, \tag{4.6b}$$

where the 'partition function' $f_j(x)$ equals

$$f_j(x) = \sum_{m=-j}^{+j} e^{(m/j)x} = \frac{\sinh{\{1+(1/2j)\}x}}{\sinh{(1/2j)x}}. \tag{4.7}$$

Equations (4·6) and (4.7) yield for $\langle\mu\rangle$ the expression

$$\langle\mu\rangle = \mu \frac{\sum\limits_{m=-j}^{+j} \frac{m}{j}\exp\left(\frac{m\mu H}{jkT}\right)}{\sum\limits_{m=-j}^{+j}\exp\left(\frac{m\mu H}{jkT}\right)}.$$

The numerator is the derivative of the denominator. We introduce therefore the abbreviation

$$L_j(x) = \frac{d}{dx}[\ln f_j(x)], \qquad (4.8)$$

where $f_j(x)$ is given by (4.7). We get then

$$\frac{\langle\mu\rangle}{\mu} = L_j\left(\frac{\mu H}{kT}\right), \qquad (4.9)$$

and

$$L_j(x) = \left(1+\frac{1}{2j}\right)\coth\left(1+\frac{1}{2j}\right)x - \frac{1}{2j}\coth\frac{1}{2j}x. \qquad (4.10)$$

There is a great family resemblance among these so-called Langevin functions $L_j(x)$.[†] They are all proportional to x for small x and equal to 1 for large x. Fig. 4.1 shows the two extreme members of the family, $j = \frac{1}{2}$ and $j = \infty$. Their behaviour for small x is of particular importance. The total magnetic moment of the sample

$$M = N\langle\mu\rangle \qquad (4.11)$$

is the magnetization, if taken per unit volume. If the low field development of (4.10) is inserted into this we get

$$\frac{M}{H} \approx \frac{1}{3}\left(1+\frac{1}{j}\right)\frac{N\mu^2}{kT}. \qquad (4.12)$$

This expression is the low field susceptibility. It is inversely proportional to the absolute temperature; this form of temperature dependence of the susceptibility is known as the *Curie law*;[‡] the law is so well obeyed by many salts at low temperature that it is

† P. Langevin, *J. de Phys.* 4 (1905), 678. P. Debye, *Handb. der Radiologie*, 6 (1925), 713.
‡ P. Curie, *Ann. Chim. Phys.*, 5 (1895), 289.

often used for temperature calibration. In the high field limit $\langle \mu \rangle$ and μ become equal. The specimen saturates, and (4.11) becomes simply

$$M \approx N\mu. \tag{4.13}$$

The specific heat behaviour of Langevin paramagnetism has certain aspects of non-co-operative ordering processes in general.

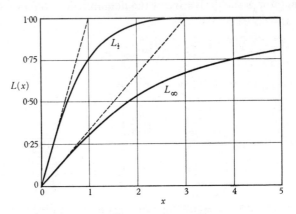

Fig. 4.1. Langevin curves giving magnetization versus magnetic field for a set of orientable dipoles.

Let us restrict discussion to the two-level case $j = \frac{1}{2}$, which corresponds to paramagnetism due to the spin of one unpaired electron. (4.10) is then specialized to

$$L_{\frac{1}{2}}(x) = \tanh x. \tag{4.14}$$

Substitution of this into (4.7) and (4.9) and differentiation with respect to T yields a paramagnetic contribution to the heat capacity at constant magnetic field equal to

$$C = \frac{N\mu^2 H^2}{kT^2}\operatorname{sech}^2\frac{\mu H}{kT}, \tag{4.15}$$

where N is the number of spins in the sample. The behaviour of C as function of T is shown in fig. 4.2. It equals zero for both high and low temperature and comes to a peak at an intermediate value. This type of specific heat curve is typical for many ordering processes: a specific heat hump, clustered around a definite temperature. In practice this hump must be assumed superimposed on a lattice

background of the type shown in fig. 3.19. There is thus a qualitative difference between the two types of curves which makes
ordering processes relatively easy to identify. If such a hump is
found empirically, and is separable from its background, the
generalization of (4.1 a) can be formed which is

$$\int_0^\infty C(T)\,dT = \Delta(U - MH),\qquad (4.16a)$$

here U is the internal energy and $-MH$ the magnetic potential
energy just discussed. The quantity in brackets on the right is
thus a magnetic form of enthalpy as indicated for (4.1a). (4.1b)
generalizes directly and yields

$$\int_0^\infty \frac{C(T)\,dT}{T} = \Delta S.\qquad (4.16b)$$

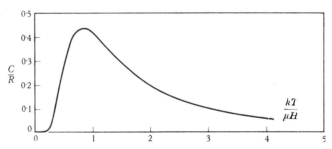

Fig. 4.2. Specific heat of a specimen containing paramagnetic ions; plot of the
magnetic contribution versus temperature at a non-vanishing field.

In the present case insertion of (4.15) into (4.16) yields

$$\Delta U = N\mu H,\qquad (4.17a)\qquad\qquad \Delta S = Nk\ln 2.\qquad (4.17b)$$

This means with (4.2) that the total number of states available to
the solid is 2^N times larger on the high than on the low temperature
side of the specific heat hump. The combination of (4.1) and (4.16)
is fairly revealing for structural transformations in general. What
is peculiar for this particular transformation, of course, is that the
temperature range in which it occurs can be shifted at will by
changing the applied magnetic field. In particular, the entropy of
the spin system of a solid can be increased by the amount (4.17b)
simply by demagnetizing a magnetically saturated specimen. If
the process is carried out adiabatically, a compensating decrease in

temperature will be enforced to keep the entropy constant. Processes of this kind are employed currently to reach the lowest temperatures known to man.

FERROMAGNETISM

In the context of this chapter we are interested in paramagnetism chiefly as a simple, well understood, non-co-operative ordering phenomenon. If we add a co-operative coupling of spins we are

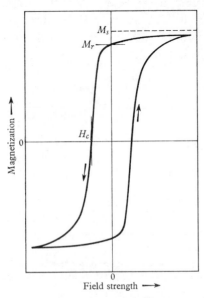

Fig. 4.3. Empirical magnetization curve of a ferromagnetic material; the curve is irreversible, and is roughly characterized by the saturation magnetization M_s, the remanence M_r and the coercive field H_c.

switching to another important field of physics, ferromagnetism. We call a material ferromagnetic if it is capable of having a magnetic moment in the absence of an applied magnetic field. Ferromagnetism is not very widespread among substances in general; the majority of ferromagnets known are alloys or compounds of the elements iron, nickel, cobalt. If a ferromagnet is heated it loses its characteristic properties at a definite temperature, the so-called *Curie temperature* T_c. Above the Curie point, ferromagnetic materials are paramagnetic, with an enhanced value of the paramagnetic susceptibility. Fig. 4.3 shows a typical plot of the magnetization as a function of an applied magnetic field for a ferromagnetic material (magnetization curve). It is seen that we deal with a phenomenon showing hysteresis; the magnetization is not a unique function of the field, but depends upon past history. In particular, the *remanence* M_r, that is the magnetization at zero field, depends on the direction in which the specimen was last saturated. Similarly, a finite opposing field, the *coercive field* H_c, is necessary to bring the magnetization to zero. A characteristic aspect is the saturation of the magnetization at a value M_s as the field grows. This aspect is reminiscent of the Langevin curves shown in fig. 4.1. However, paramagnetism saturates only at very high fields (10000 oersted) or very low temperature, while in ferromagnetism the phenomenon occurs in easily accessible ranges of temperature and field. The one element which is truly reminiscent of Langevin's theory is the value of the saturation magnetization which is essentially the same for paramagnets and ferromagnets.

The interpretation of the mechanism of ferromagnetism has long been handicapped by the obvious non-equilibrium aspect of the phenomenon as commonly observed. The phenomenon as depicted in fig. 4.3 is non-conservative; in fact, the area of the magnetization curve gives directly the amount of energy converted into heat in one hysteresis cycle. The phenomenon contains therefore frictional aspects which are not capable of a direct atomic explanation. Pierre Weiss† took the fundamental step necessary for an understanding by postulating that the atomic nature of ferromagnetism is obscured to us by a structure involving relatively large magnetized blocks, called *domains*. A single domain was supposed to have a magnetization curve of the type shown in fig. 4.4: the magnetization appears as a definite function of temperature and field. Thermodynamics and statistics are therefore applicable. What distinguishes the curve from the Langevin curve of fig. 4.1 is the *spontaneous magnetization* which the material possesses at zero field. According to Weiss those aspects of fig. 4.3 not reproduced in fig. 4.4. were to be explained by the assumption of interlocking domains which are spontaneously magnetized, but are not free to follow external changes reversibly. This hypothesis of Weiss has been brilliantly verified in the last few years. We shall not try to discuss this aspect here,

† P. Weiss, *J. de Phys.* **6** (1907), 667.

but have to refer the reader to the appropriate literature.† The atomic aspect of ferromagnetism, represented by fig. 4.4, is in itself a very interesting problem.

The original hypothesis of Weiss to explain fig. 4.4 was the hypothesis of an *internal field* proportional to the magnetization. At first sight this seems a reasonable hypothesis because such an effect is known from electrodynamics. Let us therefore modify equation (4.9) by the substitution

$$H \to H + \alpha M, \qquad (4.18)$$

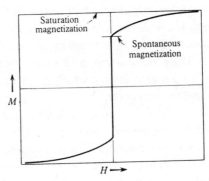

Fig. 4.4. Magnetization curve of a single magnetic domain; the curve is reversible, and differs from the Langevin curves mainly through the presence of a spontaneous magnetization at zero magnetic field.

where M is the magnetization and α an undetermined number. We get then for the case $j = \frac{1}{2}$, with (4.11) and (4.14)

$$M = N\mu \tanh \frac{\mu(H + \alpha M)}{kT}, \qquad (4.19)$$

where N is the number of spins in a unit volume.

The equation of state (4.19), connecting the thermodynamic variables M, H, T, is in qualitative agreement with the facts about ferromagnetism. First, the fact about spontaneous magnetization. Setting H in (4.19) equal to 0 and solving for T we get

$$kT = \frac{\alpha \mu M}{\operatorname{artanh}(M/N\mu)}. \qquad (4.20)$$

† See the literature list at the end of the chapter.

The equation yields a T for every M between 0 and $N\mu$. When M is almost equal to $N\mu$ the denominator is very large and T is correspondingly small. Thereupon as M decreases T increases. However, T does not increase indefinitely because the inverse hyperbolic tangent is always larger than its argument (see fig. 4.1). We therefore find

$$kT < \frac{\alpha\mu M}{M/N\mu} = \alpha N\mu^2 = kT_c. \tag{4.21}$$

When the inequality does not hold the only solution of (4.20) is $M = 0$. The temperature $T = T_c$ separates therefore a high temperature region, in which the specimen is paramagnetic, from a low temperature region, in which it is ferromagnetic. This transition temperature is the Curie temperature discussed earlier. Fig. 4.5 shows the spontaneous magnetization versus temperature curve predicted by (4.20) together with empirical results for iron, nickel, cobalt and magnetite; it is seen that the agreement is close. A particularly interesting aspect of it is the way the magnetization approaches zero near the Curie point. To see this in detail from (4.20), we may assume $M/N\mu$ small and expand the artanh, thus

$$kT \approx \frac{\alpha\mu M}{\dfrac{M}{N\mu} + \dfrac{1}{3}\left(\dfrac{M}{N\mu}\right)^3},$$

or

$$\frac{M}{N\mu} \approx 3\left(\frac{T_c}{T} - 1\right)^{\frac{1}{2}}. \tag{4.22}$$

This is an approach to zero with infinite slope. In addition to the spontaneous magnetization, the internal field theory also predicts correctly the susceptibility of ferromagnetic materials above the Curie temperature. In the paramagnetic range M is small compared to its saturation value, and therefore, the tanh in (4.19) is small and can be set equal to its argument; the resultant equation for M is, with (4.21)

$$M = \frac{N\mu^2 H}{k(T - T_c)}. \tag{4.23}$$

This equation is the *Curie–Weiss law* for the paramagnetic susceptibility of ferromagnetic materials. It differs from the previously derived Curie law only in the replacement of T by $T - T_c$. The Curie–Weiss law is also in good agreement with experiment.

It is interesting that such a successful theory as the internal field theory shows its insufficiency only when the experimental facts are used to estimate α. We get from (4.19)

$$\alpha = \frac{kT_c}{N\mu^2} \approx \frac{10^{-16} \times 10^3}{10^{23} \times 10^{-40}} = 10^4. \qquad (4.24)$$

This is in contradiction to electrodynamics which predicts α's in (4.18) at most equal to 4π. Or, to say it in another form, the observed

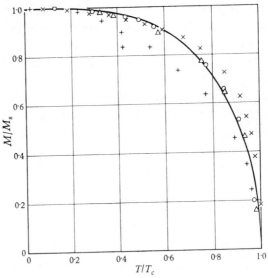

Fig. 4.5. Plot of spontaneous magnetization versus temperature; theoretical two-level internal field curve with experimental points for iron ×, nickel ○, cobalt △ and magnetite + (after R. Becker).

Curie temperatures, which lie in the neighbourhood of 1000°, are much too high if it is assumed that the atomic magnetic dipoles couple through their magnetic interaction.†

† It is interesting that in the related phenomenon of ferroelectricity, discussed below, the conclusion goes the other way. Electric moments are about two orders of magnitude larger (in Gaussian units) and ferroelectric Curie points generally lower than their magnetic analogues. The field constant α thus comes out to be of reasonable order; the internal field hypothesis is therefore right in this case, at least in its basic idea. Of course, the coupling of electric dipoles through their electrostatic interaction has also a local aspect, which is rather more complicated than the ferromagnetic coupling discussed here.

The discovery of the correct nature of the ferromagnetic coupling force came as a by-product of quantum mechanics. Dirac[†] showed that the electronic spin and the Pauli exclusion principle combine in such a way as to produce between the spins of two neighbouring electrons a coupling of the form

$$V = J \boldsymbol{\sigma}_1 \cdot \boldsymbol{\sigma}_2. \tag{4.25}$$

Here σ_1 and σ_2 are the two spins and J is a function of distance called the *exchange energy* or exchange coupling. It can be of either sign. If the wave functions of the two electrons interpenetrate substantially the exchange energy is of the same order as the electrostatic interaction, but at larger distances it falls off exponentially as the wave functions themselves. We may therefore think of J as being of electrostatic order, but only acting between close neighbours. Heisenberg[‡] was the first to realize that this exchange energy J, if of negative sign, offers a natural explanation for the phenomenon of ferromagnetism. For a local coupling of spins of the form (4.25), strung throughout a crystalline material, will tend to align spins to form a large resultant; in atoms, this effect is well known as Hund's rule and arises from just this cause.

THEORY OF THE ISING MODEL

The result of a coupling of the form (4.25), strung throughout a crystal, can be accurately evaluated only in the limit of very low and very high temperatures. The difficulties of evaluation are greatest just in the most interesting region, the region of transition from ferro- to paramagnetism. All discussions of the phenomenon are therefore forced to make use of approximations or intuitive reasoning. We shall employ here the *Ising model*[§] for the purpose of discussion. The Ising model is essentially a truncated exchange hamiltonian. It replaces $\sigma_{1x}\sigma_{2x} + \sigma_{1y}\sigma_{2y} + \sigma_{1z}\sigma_{2z}$ by $\sigma_{1z}\sigma_{2z}$ only; this is qualitatively plausible because, if the quantization is along the z direction, only this part is diagonal, and the expectation values of the other operators are zero. In addition, the Ising problem

[†] P. A. M. Dirac, *Proc. Roy. Soc.* A, **123** (1929), 714. See also any textbook on quantum mechanics.
[‡] W. Heisenberg, *Z. Phys.* **49** (1928), 619.
[§] E. Ising, *Z. Phys.* **31** (1925), 253.

needs no quantum mechanics for its discussion, because all variables of the truncated hamiltonian commute. The quantum variable
σ_z thus becomes a classical scalar σ capable of two values $+1$ or -1.
A network of such scalar spins is assumed to stretch through the
crystal, forming a coupled system. The energy U of the system is
assumed to equal

$$U = -\tfrac{1}{2} \sum_{i,k} J_{ik}\,\sigma_i\,\sigma_k - \mu H \sum_i \sigma_i, \qquad (4.26)$$

where each summation runs over all spins. The magnitude of the
interaction constants J_{ik} is almost always chosen so as to leave only
adjacent pairs in the first sum. In fact, in some forms of the model
there is just one coupling constant linking all neighbours. (4.26)
then takes the form

$$U = -J \sum_{\langle i,k \rangle} \sigma_i\,\sigma_k - \mu H \sum_i \sigma_i. \qquad (4.27)$$

Here the summation $\sum_{\langle i,k \rangle}$ runs over all pairs of direct neighbours.
The suppression of the x and y components of spin in the Ising
model produces consequences which one must keep in mind to
interpret the results of the model reasonably. In the first place, it
deprives magnetism of its angular momentum aspect and thus
falsifies its dynamics. It is, furthermore, incorrect at low temperature; at low temperature the energy is quadratic in the x and y
components of spin; the theory of small vibrations discussed in
Chapter 3 is therefore applicable and the small departures from
equilibrium take the form of *spin waves*. These spin waves are
absent in the Ising model in which equilibrium can only be dis-
turbed by actual spin reversal. It appears, however, that the
importance of these defects is relatively minor at and above the
Curie point, where the statistical count of states assumes pre-
ponderant importance; this count is right in the model. We may
therefore presume that the co-operative aspect of an exchange
coupled network is similar to the one resulting from the truncated
hamiltonians (4.26) or (4.27).

The first thing to do in a discussion of the Ising model is to
simplify it still further so as to yield an approximate derivation of
the Weiss theory. In this approximate derivation we treat the plus
spins and minus spins as two chemical species. Let N_+ be the number

of plus spins and N_- the number of minus spins, and define their concentrations by the customary relations

$$c_+ = \frac{N_+}{N_+ + N_-} = \frac{N_+}{N}, \qquad (4.28a)$$

$$c_- = \frac{N_-}{N} = 1 - c_+. \qquad (4.28b)$$

If we consider these concentrations fixed but the distribution of the spins random, then the entropy of the arrangement is given by the well-known formula for the entropy of mixture:[†]

$$S = -Nk(c_+ \ln c_+ + c_- \ln c_-). \qquad (4.29)$$

To estimate the energy, we limit ourselves to the case for which there is only one nearest neighbour interaction J. The assumption of a random arrangement of spins implies then that the relative abundances of $++$, $--$, and $+-$ neighbour pairs have their random values, that is, are as c_+^2, c_-^2 and $2c_+ c_-$ respectively. If we insert this into the energy expression (4.27) we get

$$U = -\tfrac{1}{2} z N J (c_+^2 + c_-^2 - 2c_+ c_-) - N\mu H (c_+ - c_-). \qquad (4.30)$$

Here z is the number of nearest neighbours of any given spin (six for a simple cubic lattice, eight for a body-centred cubic lattice, etc.). To simplify further, let us introduce the total magnetic moment M as a variable through the substitutions

$$c_+ = \frac{1}{2}\left(1 + \frac{M}{N\mu}\right), \qquad (4.31a)$$

$$c_- = \frac{1}{2}\left(1 - \frac{M}{N\mu}\right). \qquad (4.31b)$$

We then get for U

$$U = -\frac{1}{2}\frac{zJM^2}{N\mu^2} - MH, \qquad (4.32)$$

and for S, from (4.29),

$$S = Nk\left[\ln 2 - \frac{1}{2}\left(1 - \frac{M}{N\mu}\right)\ln\left(1 - \frac{M}{N\mu}\right)\right.$$
$$\left. - \frac{1}{2}\left(1 + \frac{M}{N\mu}\right)\ln\left(1 + \frac{M}{N\mu}\right)\right]. \qquad (4.33)$$

[†] For a derivation of this formula see any thermodynamics textbook; for instance, E. Fermi, *Thermodynamics* (New York, Prentice-Hall, 1937), equation 155, p. 116.

Now we introduce the free energy A

$$A = U - ST. \tag{4.34}$$

One of the forms of the second law of thermodynamics states that in thermodynamic equilibrium this quantity A must be a minimum for fixed temperature and external constraints. If, therefore, we succeed in approximating this quantity for a given physical situation, and if we find the approximation contains one or more undetermined parameters describing the internal state of the system, then we must dispose of them in accordance with the second law, and make A a minimum with respect to these parameters. In the present instance, we have one such parameter M; we must dispose of it by demanding that

$$\partial A / \partial M = 0. \tag{4.35}$$

This yields, after some rearrangement of terms,

$$\frac{M}{N\mu} = \tanh \frac{\mu}{kT} \left(H + \frac{zJM}{N\mu^2} \right). \tag{4.36}$$

We recognize in (4.36) the equation (4.19) previously derived, with a physical interpretation for the 'internal field constant' α:

$$\alpha = zJ/N\mu^2. \tag{4.37}$$

Since N is the number of spins per unit volume it is of the order $1/r^3$ where r is the distance between neighbouring spins; the denominator of (4.37) is thus of the order of the magnetic interaction energy μ^2/r^3 of two neighbouring spins, which is roughly 10^{-16} ergs. The numerator, on the other hand, is the exchange coupling which is more like an electrostatic quantity; this means that it may rise toward 10^{-12} ergs. The large magnitude of α is thus explained, and the 'internal field' idea reduced to a more rational set of assumptions. With the reduction of α to fundamental constants, quantities derived from it are also so reduced. We shall note here only the new expression for the Curie temperature T_c which replaces (4.21), namely

$$zJ/kT_c = 1. \tag{4.38}$$

The Curie point transition is characterized, not only by the magnetic effects just discussed, but by a characteristic thermal behaviour. This thermal behaviour is worth discussing here because it is more indicative of co-operative phenomena in general than the variation of the magnetic variables. Viewed from this point of view, paramagnetism and ferromagnetism deal with the same ordering process, first without, then with co-operative action. The modification of the thermal behaviour as a consequence of this action is then of basic interest. Part of this modification can be predicted from our general equations (4.16). The coupling makes the energy change larger, thereby increasing the heat capacity so as to increase (4.16a). At the same time, however, the expression (4.16b) stays the same. This can be accomplished only by shifting

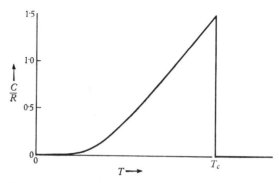

Fig. 4.6. Specific heat versus temperature in the internal field theory of ferromagnetism; comparison of this curve with fig. 4.2 shows the effect of a co-operative coupling.

the entire transformation to a higher temperature. However, if these were the only aspects we probably would not pay much attention to co-operative change. Fig. 4.6 shows the specific heat curve of the internal field model for zero field. It is obtained by computation of dU/dM from (4.32), dM/dT from (4.20), and multiplication of the two. What is most striking about the curve is not the shift in temperature, but the increase in abruptness of the curve: the specific heat rises to a maximum and thereupon goes discontinuously to zero. Yet at the same time, the change is not as abrupt as for normal phase changes, where two forms of different material properties abut at one temperature. Transitions of this type are called second order phase changes.

The question arises whether we can account for such transitions from fundamental theory. At first sight one might be tempted to answer this question in the affirmative, since we have discussed all along theoretically derived equations. However, the internal field theory proceeds from an unphysical assumption, and for the Ising model the derivation is only approximate. Indeed, many different ways of deriving the properties of the Ising model have been devised, and each one of them gives rise to a different answer. The problem of evaluating the results of the model correctly has therefore major importance.

Careful statistical evaluations of a model such as the Ising model proceed almost always through the partition function f which is defined as

$$f = \sum_{\nu} e^{-E_{\nu}/kT},$$

where the index ν runs over all states of the system and E_{ν} is the energy of these states. In the present instance the energy is given by (4.26) or (4.27), and all possible states of the system are exhausted by summing over all possible combinations of values ± 1 of the Ising spins. This yields

$$f = \sum_{\sigma_i = \pm 1} \exp\left[\frac{J}{kT}\sum_{\langle i,k \rangle} \sigma_i \sigma_k + \frac{\mu H}{kT}\sum_i \sigma_i\right]. \qquad (4.39)$$

Here the sum $\sum_i$ goes over all spins, $\sum_{\langle i,k \rangle}$ over all pairs of direct neighbours, and $\sum_{\sigma_i = \pm 1}$ over the 2^N combinations ± 1 of the N spins. The evaluation of f is sufficient to yield knowledge of all thermodynamic quantities. In particular, we get for the energy U of the system

$$U = kT^2 \frac{\partial \ln f}{\partial T} \qquad (4.40)$$

and for the total magnetic moment M

$$M = kT \frac{\partial \ln f}{\partial H}. \qquad (4.41)$$

The study of f for cases of physical interest is an extremely hard problem. Its difficulty resides in the fact that, if one wishes to evaluate it for the purpose of investigating phase transitions, one cannot assume the number of co-operating units finite; for in this latter case, (4.39) consists of a finite sum of terms without temperature singularity; this makes the sum likewise non-singular. All interesting results are therefore obtained by examination of the Nth root of f in the limit that N goes to infinity. The difficulties of a limiting process are therefore superimposed upon the combinatorial difficulty of evaluating (4.39).

Some further thought along the same lines shows that even a one-dimensionally infinite system cannot show ferromagnetism. For ferromagnetism means a spontaneous magnetization, which in turn means long-range order among the spins without an applied field. Suppose that we have such an ordered arrangement of N Ising spins, with all spins parallel. Then one can upset it by introducing a single break in it, that is, introducing the energy $2J$ at

$$+ \; + \; + \; + \; + \; + \; + \; + \; + \; + \; + \; +$$

$$+ \; + \; + \; + \; + \; + \; - \; - \; - \; - \; - \; -$$

Fig. 4.7. Destruction of order in a linear array of spins; one single reverse coupling is sufficient to destroy long-range order.

one point, as shown in fig. 4.7. Since this break can be made at any of N positions (where $N+1$ is the number of members of the linear chain) the entropy gain is $k \ln N$. The free energy change is therefore
$$\Delta A = 2J - kT \ln N.$$

Since we have to consider ΔA in the limit of infinite N there is no temperature, however small, sufficient to prevent destruction of long range order, because A is always lower for the disordered state.

The same argument was developed by Peierls† to prove the contrary for a square net of Ising spins: such a net is ferromagnetic. For this purpose an estimate must be made of the number of ways a border of L segments, separating $+$ and $-$ spins can be laid in a square net. Starting from one point to lay out such a border, one can at every step except the first proceed in at most three different directions (see fig. 4.8). This gives at most $4 \times 3^{L-1}$ different ways of laying out such a border, starting from a fixed point. The energy necessary to lay it is $2JL$. For geometrical reasons, no border can

† R. Peierls, *Proc. Camb. Phil. Soc.* **32** (1936), 477.

be less than four units long. Hence the probability that there is a border of length L starting at a point is

$$\frac{4 \times 3^{L-1} e^{-2JL/kT}}{1 + \sum\limits_{L=4}^{\infty} 4 \times 3^{L-1} e^{-2JL/kT}} \quad (L \geqslant 4).$$

We overestimate all numbers if we leave the sum off in the denominator; the probability is therefore less than

$$4 \times 3^{L-1} e^{-2JL/kT}.$$

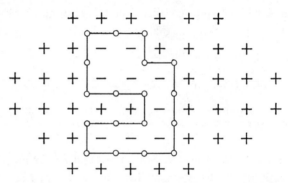

Fig. 4.8. Introduction of reverse orientation into the square Ising net; it is now necessary to maintain a border of high energy.

A border of length L can be shifted parallel to itself into N/L non-equivalent positions. This yields an expectation value of at most

$$4 \frac{N}{L} 3^{L-1} e^{-2JL/kT}$$

for the number of such borders. We now try to get an upper limit for the area such a border encloses by taking it as $\frac{1}{16}L^2$. The total area whose spin could be reversed by all possible borders is then at most the sum of the product of the last two numbers

$$A < \sum_{L=4}^{\infty} \frac{1}{16}L^2 \times 4 \frac{N}{L} 3^{L-1} e^{-2JL/kT}.$$

The expression converges as soon as $\kappa = 3 e^{-2J/kT}$ is less than 1. It becomes then

$$A < \frac{1}{12}N \frac{4\kappa^4 - 3\kappa^5}{(1-\kappa)^2}.$$

For $$\kappa < \tfrac{1}{2}, \quad J/kT > 0.9,$$

the maximum number of reversed spins is thus smaller than $\tfrac{1}{12}N$, and the sample has a resultant moment. Thus there is a temperature below which boundaries cost too much in free energy to prevent spontaneous magnetization. Perusal of this argument shows that it rests on the observation that energy and entropy both vary as the first power of L, and that therefore a temperature can be found for which the energy overpowers the entropy in determining the equilibrium state of the Ising net.

It has proved possible to go beyond estimates such as the one given; we can prove not only that the square Ising net has a spontaneous magnetization at a low enough temperature, but we can derive for it and some related models its thermodynamic properties and even many correlation quantities. All models for which this has been achieved are at most two dimensionally infinite. In fact it appears that any reasonably simple regular two-dimensional net without cross-overs can be analysed for its thermal properties and spontaneous magnetization. Among the methods used the only one having general implications is one which transforms the problem of summing over spins extending over n infinite dimensions to an eigenvalue problem involving spins covering only $n-1$ infinite dimensions. From what was said, this reduction is not enough to get at the important results. Recourse must therefore be had to other procedures; these procedures are very hard, not directly accessible to physical intuition, and adapted only to the Ising net in two dimensions.† We shall therefore omit derivations and proceed to examine results only. Let us choose the rectangular Ising net as a typical example of fair generality.

The rectangular Ising net is analogous to the square net discussed above but has two interactions J and J' coupling at right angles to each other (fig. 4.9). The significant quantity is the limiting expression

$$\lambda = \lim_{N=\infty} [f(N)]^{1/N}, \tag{4.42}$$

where $f(N)$ is given by (4.39). It is seen from (4.40) and (4.41) that such a λ will give specific properties per spin of the entire net. In

† For details, see the review article of Newell and Montroll quoted at the end of the chapter.

letting N go to infinity we must of course take care to proceed to infinity for both dimensions of the two-dimensional net. As a result of the calculations mentioned it is found that the λ for the rectangular net equals

$$\ln \lambda = \tfrac{1}{2}\ln\{4\cosh(2J/kT)\cosh(2J'/kT)\} +$$

$$\frac{1}{2}\int_0^1\int_0^1\ln\left\{1 - \frac{\sinh(2J/kT)\cos 2\pi f + \sinh(2J'/kT)\cos 2\pi f'}{\cosh(2J/kT)\cosh(2J'/kT)}\right\}\,df\,df'. \tag{4.43}$$

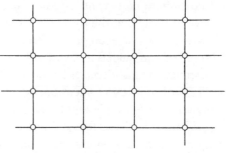

Fig. 4.9. The rectangular Ising net; this form of the Ising model has two coupling constants, one along the short sides and one along the long sides.

For the particular case of the square net, $J = J'$, the expression for λ can be given the shorter form

$$\ln \lambda = \ln\{2\cosh(2J/kT)\}$$

$$+ \frac{1}{\pi}\int_0^{\frac{1}{2}\pi}\ln\left\{\frac{1+(1-\kappa^2\sin^2\phi)^{\frac{1}{2}}}{2}\right\}\,d\phi, \tag{4.44}$$

where the auxiliary parameter κ is defined through the relation

$$\kappa = \frac{2\sinh(2J/kT)}{\cosh^2(2J/kT)}. \tag{4.45}$$

The expressions (4.43) and (4.44) have a form typical of all the results obtained. A first term, elementary in form, gives a rough outline of the behaviour of λ as function of the temperature. For instance, if we insert the first term of (4.44) into (4.40) to compute the energy, we find

$$U \approx -2NJ\tanh\frac{2J}{kT}. \tag{4.46}$$

This result is correct in the limit of very high or very low temperature, but varies smoothly with the temperature: it is in fact the result for a linear Ising chain with coupling $2J$. Added to this term, however, there is another which has the form of a trigonometric integral; this added term vanishes at very low and very high temperature but rises to a peak at intermediate temperature in such a way as to produce a profound modification of the result (4.46). At the place at which the second term reaches its maximum it also becomes singular. Let us follow this in detail for the formula (4.44). It follows from the properties of the hyperbolic functions that the parameter κ defined in (4.45) is never larger than 1. Its maximum is reached at the value T_c for which

$$\sinh \frac{2J}{kT_c} = 1, \qquad (4.47a)$$

or
$$\frac{J}{kT_c} = 0{\cdot}4407. \qquad (4.47b)$$

At this position the integrand of (4.44) becomes singular and confers this property upon λ. The nature of the singularity is seen more clearly if the energy is computed with the help of (4.40) and (4.42). One finds
$$U(T) = -NJ \coth \frac{2J}{kT}\left[1 \pm \frac{2}{\pi}(1-\kappa^2)^{\frac{1}{2}} K(\kappa)\right], \qquad (4.48)$$

the + sign holding below the Curie point and the − sign above. κ is the parameter defined by (4.45) and $K(\kappa)$ the complete elliptic integral of the first kind. The singularity is contained in the second term of the square bracket. It equals zero at the Curie point and behaves as $-(T-T_c)\ln|T-T_c|$ in its immediate neighbourhood, the logarithm arising from the logarithmic singularity contained in $K(\kappa)$. From this we find by differentiation that the specific heat is logarithmically infinite at the critical point.

Essentially the same type of results arise in all two-dimensional cases studied. For example, analysis of equation (4.43) leads to a unique singular temperature T_c which is given by the formula

$$\sinh \frac{2J}{kT_c} \sinh \frac{2J'}{kT_c} = 1. \qquad (4.49)$$

The nature of the singularity is the same as in the case $J = J'$.

Fig. 4.10 shows the specific heat curves for the cases $J' = J$, $J' = \frac{1}{100}J$ and $J' = 0$.

In fig. 4.11 the result of equation (4.48) is shown in heavy outline in comparison with the internal field result in light outline; this latter result is obtained by elimination of M from (4.32) and (4.36); several other proposed approximate solutions are added in dashed outline. It is seen that while the internal field approximation, as

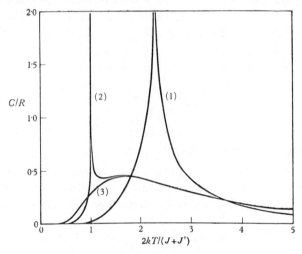

Fig. 4.10. Specific heat versus temperature curves for the rectangular Ising net. (1) $J = J'$, (2) $J = \frac{1}{100}J'$, (3) $J = 0$ (after Onsager).

well as several others, suggests the actual behaviour of the model, it departs from the correct answer in its details: location and nature of the temperature singularity. It is also seen that improvement of the approximation reduces the quantitative discrepancies, but not the qualitative errors.

The situation just described also exists with respect to the magnetic properties of the Ising model. Yang† has calculated the spontaneous magnetic moment M of the square Ising net as a function of temperature; it is zero above the Curie point and below it has the value

$$M = N\mu \left[\frac{\cosh^2(2J/kT)}{\sinh^4(2J/kT)} \{\sinh^2(2J/kT) - 1\} \right]^{\frac{1}{8}}. \quad (4.50)$$

† C. N. Yang, *Phys. Rev.* **85** (1952), 809.

Its behaviour is shown in fig. 4.12. The hope that the two-dimensional Ising net would provide a mechanically consistent model of ferromagnetism is thus confirmed. In detail there are discrepancies with experiment. The vanishing of the magnetization with the $\frac{1}{8}$th power of $T_c - T$ agrees neither with the internal field picture, nor with experiment, as is evident from fig. 4.5. Similarly, the specific heat curves of fig. 4.10 show a singularity which is about the same on the two sides of the singular point. In fact, one can go further,

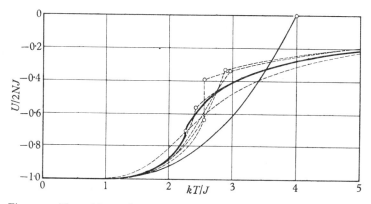

Fig. 4.11. Plots of internal energy versus temperature for the square Ising net. The exact curve is shown in heavy outline and the internal field approximation in light outline. Other proposed approximations are shown dashed.

and prove directly from a topological-algebraic argument that all simple two-dimensional Ising models have a symmetry between high and low temperatures. In the case of equation (4.44), this shows up in the dependence of the result on the parameter κ defined in (4.45), which has the same value for a pair of temperatures, one above, and one below T_c.

The properties of the three-dimensional Ising lattices are not known accurately at this time. We cannot say, therefore, whether their properties agree with experiment in detail. There is some evidence in that direction from numerical work. However, there is also evidence against hoping for a good fit; this evidence is based on inadequacies of the Ising approximation as compared to the complete interaction (4.25). An instance of the distinction between these two interactions is the fact that the proof of Peierls given above breaks down for the interaction (4.25). Indeed, if the interaction is given by (4.25), the energy due to a pair of neighbours is proportional to $1 - \cos \vartheta$, where ϑ is the angle between their spins. This opens up the possibility of reducing the energy of a border such as the one shown in fig. 4.8 by insertion of a transition layer in which spins reverse their direction gradually. If the reversal

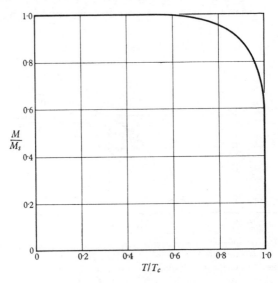

Fig. 4.12. Plot of the spontaneous magnetization versus temperature for the square Ising net; observe how the curve differs from the curve of fig. 4.5.

is accomplished in m steps the energy of each step is reduced by $1/m^2$, while the number of steps is only increased by m. The energy thus carries a factor $1/m$. We can now remove the length factor L altogether in the energy of a domain boundary by choosing m of the order L; the energy of a reversed domain thus becomes independent of its size, as for the Ising model in one dimension. This change collapses the proof, and permits, in fact, proof of the opposite proposition. In practice the Ising model is not quite as wrong as these thoughts indicate, because even slight anisotropy favouring the Ising direction will prevent the domain boundary from getting too thick and will restore thereby the validity of the Peierls proof. Such anisotropy is always present in actual domain boundaries. Anisotropy and exchange coupling do in fact determine together the thickness and surface tension of such boundaries, as was shown by F. Bloch.†

ANTIFERROMAGNETISM

The discovery that many materials are antiferromagnetic is relatively recent. Antiferromagnetism arises when the exchange interaction constant in (4.25) is positive between neighbours,

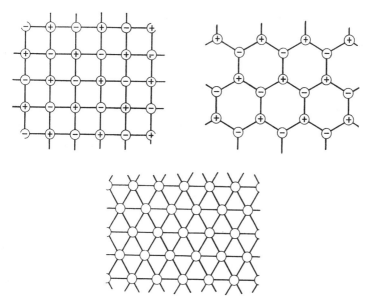

Fig. 4.13. Antiferromagnetism; subdivision of the square and honeycomb nets into subnets of opposite orientation; no such subdivision is possible for the triangular net because every elementary triangle must hold at least two neighbours of equal orientation.

tending to align them antiparallel. The simplest circumstance is one in which the entire lattice can be divided into two equivalent sublattices 1 and 2 in such a way that all nearest neighbours of a member of 1 belong to sublattice 2 and vice versa. Fig. 4.13 shows how this subdivision is made in the square and honeycomb nets, and that such an arrangement is impossible for the triangular net.

† F. Bloch, Z. Phys. **74** (1932), 295. Consult also the literature list at the end of the chapter.

The theory envisages the antiferromagnetic ground state as one in which each sublattice is saturated, but the two magnetic moments cancel. An approach to antiferromagnetism from the fundamental theory of electron spin does not justify this naïve viewpoint, because a state of total spin zero has annoying complexities, which do not exist for the state of maximum spin.† However, accidental reasons, such as strong anisotropy forces, stabilize the state against fluctuations in many cases. We shall therefore follow the simple form of the theory from now on.

For an approach to the thermodynamics of antiferromagnetism it is useful to employ again the Ising model. It gives us immediately an extremely attractive result for the thermal effects of antiferromagnetism. If the lattice can be divided into two sublattices as stated above, a one-to-one correspondence can be established between ferromagnetic and antiferromagnetic arrangements having the same absolute value of J: take an arrangement of the ferromagnet, reverse all spins of the sublattice 2, and you have an arrangement of equal energy in the antiferromagnet. Hence the specific heat will be the same, and all the discussion expended on it previously is applicable again. The theory of the spontaneous magnetization can be taken over if we introduce an 'antiferromagnetic moment'

$$P = M_1 - M_2, \tag{4.51}$$

where M_1 and M_2 are the magnetic moments of the two sublattices. There will be a transition temperature T_c below which P will have a non-zero value. Its exact behaviour is subject to the uncertainties described earlier; in the internal field approximation it will follow the curve shown in fig. 4.5. Proceeding to the magnetic susceptibility we enter new ground; for the exchange coupling is now opposed to the field, and the result must be a paramagnetic susceptibility smaller than the value (4.12). It will be observed that no results have been forthcoming concerning susceptibilities in the rigorous calculations on the Ising model; we shall therefore fall back on the internal field approximation and adapt it to our new purposes.

The logical generalization of (4.19) is due to Néel;‡ he wrote

† Lamek Hulthen, *Ark. Mat. Astr. Fys.* **26**A, no. 1 (1938). P. W. Anderson, *Phys. Rev.* **86** (1952), 694.
‡ L. Néel, *Ann. de Phys.*, Paris, **17** (1932), 64.

down two equations, one for each sublattice, and assumed the internal field for one entirely controlled by the other

$$M_1 = \tfrac{1}{2}N\mu \tanh \frac{\mu}{kT}(H - 2\alpha M_2), \qquad (4.52a)$$

$$M_2 = \tfrac{1}{2}N\mu \tanh \frac{\mu}{kT}(H - 2\alpha M_1). \qquad (4.52b)$$

Here α has the usual meaning (4.37), except that its change of sign has been introduced explicitly into the equations.† In the absence of a field H these equations are simply solved by

$$M_1 = -M_2 = \tfrac{1}{2}P, \qquad (4.53)$$

and the moment P is then seen to obey the previous equation for M. If a field is present we employ (4.51) to write

$$M_1 = \tfrac{1}{2}(P + M), \qquad (4.54a)$$

$$M_2 = \tfrac{1}{2}(-P + M). \qquad (4.54b)$$

Here M is the true magnetic moment obtained by summing M_1 and M_2. This quantity M as well as H can be treated as small since the magnetization is bound to be less even than that for a paramagnetic sample. Therefore, when substituting (4.54) into (4.52), we can safely expand to linear terms only in H and M. This yields

$$P = N\mu \tanh \frac{\mu \alpha P}{kT}, \qquad (4.55)$$

$$M = \frac{N\mu^2}{kT}(H - \alpha M)\operatorname{sech}^2 \frac{\mu \alpha P}{kT}. \qquad (4.56)$$

Equation (4.55) is identical with (4.19); it simply confirms the spontaneous value of P to first order in H. Equation (4.56) is preferably discussed in two stages. Above the transition temperature T_c, the solution P of (4.55) is zero. The sech^2 in (4.56) then equals unity and we get with the definition (4.21)

$$M = \frac{N\mu^2}{k(T + T_c)} H. \qquad (4.57)$$

† The factor 2 in front of the α is needed so that the internal field has the same value as previously for the same value of the atomic moments. Under these conditions, M_1 and M_2, being moments of sublattices, have only half the value of M in equation (4.19).

This is *Néel's law* for the susceptibility of an antiferromagnetic material; it can be considered, in a sense, the analytic continuation of the Curie–Weiss law (4.23) for a reversed sign of the coupling. As anticipated, the value of M resulting from Néel's law is indeed smaller than the paramagnetic value which corresponds to $T_c = 0$. The second stage of discussing (4.56) applies to the region below the antiferromagnetic Curie temperature T_c when P has a spontaneous non-zero value. We may write then

$$M = \frac{N\mu^2}{k(T^+ + T_c)} H, \qquad (4.58a)$$

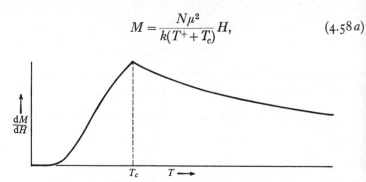

Fig. 4.14. Magnetic susceptibility versus temperature for an antiferromagnetic material; result of the two-level internal field approximation.

where T^+ is an 'equivalent temperature' given by

$$T^+ = T \cosh^2 \frac{\mu\alpha P}{kT}. \qquad (4.58b)$$

The dependence of the $\cosh^2$ on T is so strong that when T decreases T^+ will actually increase. Correspondingly M in (4.58a) will drop with the temperature rather than continue to rise. Finally at $T = 0$ the magnetic susceptibility will have become zero, quenched by the strong antiferromagnetic coupling.† The behaviour just described is shown in fig. 4.14. It is in qualitative agreement with experiment.

---

† In attenuation of this result it should be noted that the susceptibility derived here is the one parallel to P which can only be observed under special conditions. The susceptibility at right angles to P remains finite; antiferromagnetic materials tend to orient their spontaneous P at right angles to H, so that the transverse susceptibility is the one normally observed.

HIGH-LOW TRANSITIONS AND FERROELECTRIC TRANSITIONS

The Ising model also helps in the understanding of a number of co-operative non-magnetic phenomena in solids. Let us discuss a widespread type of geometrical transformation, called *high-low*

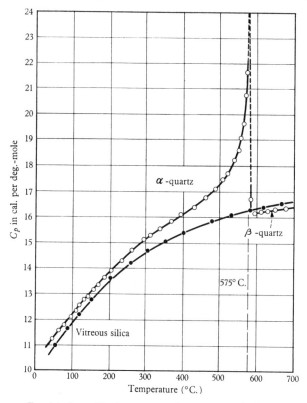

Fig. 4.15. Empirical specific heat curves for quartz and vitreous silica; the transition of quartz from the α to the β form appears to be a co-operative one; no trace of it is discernible in vitreous silica (After H. Moser).

transition. Quartz may serve as an example for the discussion of this type of phenomenon. Fig. 4.15 shows a plot of specific heat versus temperature for quartz as compared to vitreous silica. Vitreous silica has the kind of specific heat curve discussed in Chapter 3: it is gradually rising and approaching asymptotically the Dulong–Petit value $9R$. The deviation which quartz shows from this trend is typical of a co-operative phenomenon with second order phase change. Careful observation seems to exclude a latent heat. This thermal behaviour is associated with a geometrical modification of the crystal. The kind of quartz discussed in Chapter 1 and exhibited in fig. 1.42 is the high or β form. As its temperature is lowered through the transformation range it changes into an α form, in which the silicon atoms are shifted as indicated in fig. 4.16. The crystal is thereby reduced from sixfold to threefold symmetry.

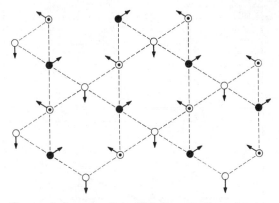

Fig. 4.16. Geometrical nature of the high-low transformation in quartz. The silicon atoms of the β structure are shown as in fig. 1.42; the arrows indicate the displacements which bring it into the α form.

In analysing this phenomenon we may start out from the idea that the silicon atoms co-operate in choosing their positions. The α positions indicated in fig. 4.16, which we call α_1, are obviously stabilized energetically through mutual interaction. However, the quartz could have gone into the α form in a second way, by having each silicon atom move in the opposite direction to the one shown in fig. 4.16. This form is rotated with respect to the first one; we call it α_2; for symmetry reasons this form must be energetically just as favourable.† We may therefore think of the configurational energy as having the appearance of fig. 4.17, with the symmetric position of fig. 1.42 as an unstable maximum.

In reducing the phenomenon to its simplest formulation we may

† It is actually found empirically that β quartz tends to break up into a mixture of α_1 and α_2 blocks when cooled. Special precautions are needed to prevent it.

presume that the transformation, which in principle could be smooth, proceeds at such high speed in a narrow temperature range because there is a built-in avalanche effect. If all neighbours are in α_1 positions, the forces exerted upon a particular atom favour for it this same position. Inversely, once the atoms of a region have random positions, the forces upon any individual atom no longer favour the arrangement α_1 or α_2. A mathematical picture of this state of affairs is gained if we associate with each silicon atom a 'spin' σ, to have the value $+1$ if the atom is in the α_1 position, and

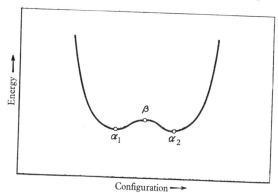

Fig. 4.17. Schematic energy plot in configuration space showing the energy of β quartz in comparison to the energy of the two α twins into which it can deform.

-1 if it is in the α_2 position, while the hexagonal β position will be assumed realized only as a dynamical average of α_1 and α_2. Then, if we assume that the coupling favouring one position over the other is between neighbours only, we can write an effective energy term of the form

$$U = -J \sum_{\langle i,k \rangle} \sigma_i \sigma_k.$$

This is exactly the equation (4.27) above, which characterizes the Ising model without a magnetic field. A confirmation of this viewpoint is furnished by experiment: the entropy change of quartz through the transformation can be estimated very roughly from Fig. 4.15 and it agrees with the theoretical prediction $R \ln 2$ which results from our reduction to the Ising model. It is clear that the distinguishing feature of high-low transitions, as compared to structural transformations in general, is the smallness of the geometrical displacements. This permits the transformation to proceed in equilibrium without destruction of the crystal matrix; the nature of the intermediate steps in the transformation can thereby be studied more easily.

It appears that *ferroelectric transformations* are a particular type of high-low transformation in which the low form is a distortion of the high form such as to possess a spontaneous dipole moment. Thus ferroelectrics are borderline cases of crystals which possess a structurally permanent electric moment. Crystals which possess permanent moments are usually compensated by external charges, and observation of their moment must be made by indirect means: in *electrets* the polar moment is constantly restored by evaporation of the compensating surface layer, in *pyroelectrics* the change of the moment with temperature can be observed because the charge compensation is too slow to follow the heating cycle. Compared to these substances with built-in moment, ferroelectrics are marginal, because the form with a moment is distinguished from the one without only by a small deformation. Like a high-low deformation, this deformation comes and goes reversibly with temperature; what is special for the ferroelectrics is that the low form has an electric moment and is thus influenced by the conjugate variable of the moment, an applied electric field.

Not all ferroelectrics are sufficiently simple to allow a semi-quantitative follow-up of these ideas. A favourable case is KH_2PO_4, potassium dihydrogen phosphate. The high form of this crystal is tetragonal. The tetragonal axis is the ferroelectric axis of the low form. The structure sensitive elements seem to be the PO_4 tetrahedra whose arrangement with respect to the tetragonal plane is shown in fig. 4.18. In the non-polar state the upper and lower edges of these tetrahedra are equivalent and therefore of equal length. In the polar state the polarity is either out of or into the paper consequently the upper and lower edges of these tetrahedra are no longer equivalent, and they will acquire different lengths; the polar form is therefore orthorhombic. It is an implicit consequence of this analysis, that a shear in the x-y-plane and a field along the z-axis are equivalent in their influence on the ferroelectric transformation; a second implicit consequence is that the entropy of the transformation is about $R \ln 2$. Both these consequences are confirmed by experience; the description of such a PO_4 group by an Ising spin is therefore appropriate. However, the second aspect of ferroelectricity, the coupling of the spins, does not fit so well into an Ising picture. The coupling arises through dipolar forces which fall off relatively slowly with distance. The nearest-neighbour picture is therefore poorer than for ferromagnetism;

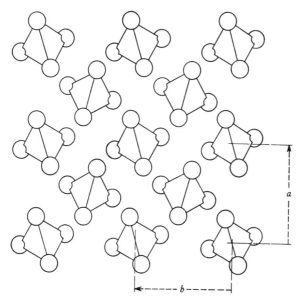

Fig. 4.18. Arrangement of the PO_4 tetrahedra of potassium dihydrogen phosphate with respect to the tetragonal basal plane. The edges shown are in the plane and are offset slightly with respect to the direction **a**, while the far edges (not shown) follow the direction **b**; a moment perpendicular to the paper thus makes **a** and **b** non-equivalent (after Shirane, Jona and Pepinsky).

also the dipolar field has a strong angular dependence. It is therefore more natural to handle these interactions by the polarization concept and the internal field associated with it than by a nearest-neighbour model. It should be added here that even for the high-low transformation of quartz and other silicates there is no direct evidence that the nearest-neighbour picture is satisfactory. The forces are of a higher polarity than dipolar, and therefore of shorter range, so the picture is presumably better. In any case, the Ising spin helps to clarify the nature of the variable which is subject to

9   w s t

order; it shows incidentally that the transition of ferroelectrics at the ferroelectric Curie point is a special kind of high-low transition.

SUPERSTRUCTURE IN ALLOYS

Another co-operative phenomenon reducible to the Ising model is the appearance of superstructure in some alloys. Alloys are substances consisting of two or more metals; it will be shown in Chapter 8 that their bonds must be considered as chemical even though a particular bond has not a specific valency character. Nevertheless, a residual valence sometimes survives as a fringe effect. It shows itself in a tendency toward an ordered arrangement of components, if the temperature is lowered gently enough to permit these weak forces to effect the necessary transport of matter. The most clear-cut way to discover the existence of superstructure is by the appearance of superstructure lines in the X-ray diffraction pattern. There are also thermal effects; for instance, β-brass exhibits a very high asymmetric peak in the specific heat similar to the one exhibited in fig. 4.15. β-brass is a body-centred cubic alloy having the chemical composition CuZn. The ordered variant of it is the caesium chloride structure shown in fig. 1.37.

The application of the Ising model to superstructure is very direct. Let us take β-brass as an example. We assign to each lattice site the 'spin' $+1$ if it is occupied by a Cu atom and -1 if occupied by a Zn atom. The phenomenon of superstructure formation is then equivalent to a tendency toward antiferromagnetism among a set of spins capable of values ± 1. If, in addition, we make use of the valency idea to assume that ordering takes place exclusively through nearest-neighbour interaction then the theoretical problem has become the Ising problem: the only parameter is the energy $\pm J$ associated with a bond according to whether the two nearest neighbours associated with this bond are like or unlike. This mathematical identification associates the total magnetic moment of the Ising model with the composition of the alloy. The magnetization of the model thus becomes a constant which cannot be altered except by switching specimens. Statistical mechanics takes care of this situation by formulating the problem through the grand partition function, which considers all possible compositions but keeps the chemical potential of the components fixed. These potentials are then adjusted at the end of the calculation to reproduce the true composition.† To get the thermal properties of an alloy of fixed composition we have thus to study the Ising model of antiferromagnetism at variable field and fixed magnetization. We shall not follow up this modification. Suffice it to say that for our example of CuZn these modifications are not necessary. The antiferromagnetic Ising model in the absence of a field has automatically no magnetic moment. Since this corresponds to an equal mixture of Cu and Zn, the results of this calculation, and the computed specific heat in particular, are directly applicable to the alloy.

THE LATTICE GAS

A fundamental advance in our understanding of phase transitions was achieved when Yang and Lee proved that the process of sublimation of a solid into a gas can be schematized in a mechanically consistent way so as to become computable from the Ising model.‡ The argument runs as follows. The classical partition function for a gas of n particles equals

$$f_n = \frac{1}{h^{3n}} \int\int \dots \int d\mathbf{p}_1 d\mathbf{p}_2 \dots d\mathbf{x}_n \, e^{-\mathscr{H}/kT}; \qquad (4.59)$$

here $\mathscr{H}$ is the total hamiltonian of n particles. The integrations over $\mathbf{p}_i$ are supposed unlimited and those over $\mathbf{x}_i$ are supposed to go over a fixed 'box' volume. It is well known that (4.59) can be partly integrated, yielding

$$f_n = \left(\frac{2\pi mkT}{h^2}\right)^{\frac{3}{2}n} Q_n, \qquad (4.60)$$

where $$Q_n = \int\int\int \dots \int d\mathbf{x}_1 d\mathbf{x}_2 \dots d\mathbf{x}_n \, e^{-\phi/kT}; \qquad (4.61)$$

ϕ is the total potential energy of the system. From f_n the grand partition function Z can be constructed by the rule

$$Z = \sum_{n=0}^{\infty} f_n e^{\zeta n}. \qquad (4.62)$$

† This procedure is usually, but not always, equivalent to the straightforward procedure of carrying out the statistics at fixed composition. For an interesting instance of disagreement, see M. Lax, *Phys. Rev.* **97** (1955), 1419.

‡ C. M. Yang and T. D. Lee, *Phys. Rev.* **87** (1952), 404.

ζ is essentially the chemical potential of the gas. It is determined by the condition

$$N = \frac{\partial \ln Z}{\partial \zeta}. \qquad (4.63)$$

The further relation

$$U = kT^2 \frac{\partial \ln Z}{\partial T} \qquad (4.64)$$

fixes the nature of Z by thermodynamics; it equals

$$Z = e^{pV/kT}. \qquad (4.65)$$

The cross-connexion from these standard concepts of the theory of fluids to the Ising model was achieved through the concept of the lattice gas. Suppose we divide space into identical cells and schematize thereupon the interaction of two particles in the following way:

(a) The interaction of two particles shall be zero unless they are in identical or adjoining cells.

(b) The interaction energy of two particles in adjoining cells shall be $-4J$.

(c) The interaction energy of two particles in the same cell shall be $+\infty$; this means that two particles will not be allowed in the same cell.

These specifications reduce the computation of Q_n from (4.61) to a consideration of cells. Each cell can be in one of two states only; either it has, or it has not a particle in it. Let the former state be described by $\sigma = +1$, the latter by $\sigma = -1$. Then we see that combinatorially the problem is of the Ising form; the form of the interaction also has the Ising form, because the total potential energy ϕ equals

$$\phi = -J \sum_{\langle i,k \rangle} (1 + \sigma_i)(1 + \sigma_k). \qquad (4.66)$$

This is a special case of (4.27), arranged so that each term vanishes unless both spins are $+1$. With formula (4.66), Q_n as defined by (4.61), becomes

$$Q_n = v^n \sum_{\sigma_i = \pm 1}^{(n)} \exp\left[\frac{J}{kT} \sum_{\langle i,k \rangle} (1 + \sigma_i)(1 + \sigma_k) \right]. \qquad (4.67)$$

Here v is the volume of a cell and $\sum_{\sigma_i = \pm 1}^{(n)}$ is the summation over spins familiar from equation (4.39), *except that* the total number of plus spins is fixed at n, and of minus spins at $(V/v) - n$. From this, equation (4.62) takes the form

$$Z = \sum_{n=0}^{\infty} \left(\frac{2\pi mkT}{h^2}\right)^{\frac{3}{2}n} e^{\zeta n} v^n \sum_{\sigma_i = \pm 1}^{(n)} \exp\left[\frac{J}{kT} \sum_{\langle i, k \rangle} (1 + \sigma_i)(1 + \sigma_k)\right].$$

(4.68)

Obviously the restriction upon $\sum\limits_{\sigma_i = \pm 1}$ and the subsequent summation over n cancel each other. To make the cancellation explicit we use the substitution

$$n = \sum_i \tfrac{1}{2}(1 + \sigma_i). \tag{4.69}$$

This yields for Z

$$Z = e^{-zJV/2kTv} e^{\frac{1}{2}\eta V/v} \sum_{\sigma_i = \pm 1} \exp\left[\frac{J}{kT} \sum_{\langle i, k \rangle} \sigma_i \sigma_k + \tfrac{1}{2}\eta \sum_i \sigma_i\right], \quad (4.70)$$

where η is a new variable replacing ζ; it has the form

$$\eta = \zeta + \psi(T). \tag{4.71a}$$

The actual expression for $\psi(T)$ is important only if we want to calculate the energy, for which purpose (4.64) has to be modified. (4.63) remains, however:

$$N = \frac{\partial \ln Z}{\partial \eta}. \tag{4.71b}$$

With equation (4.70) the partition function of the lattice gas is reduced to that of the Ising model. It is particularly instructive to follow the character of the isotherms of the model. Varying the pressure at constant temperature means, because of (4.65), varying Z without varying T, V, v or J. This means varying η. In magnetic language, which is possible by identification of (4.70) and (4.39), this means varying the magnetic field. Now the Ising magnetization curve has the qualitative aspect of fig. 4.4, if the temperature is below the Curie point. It consists of three pieces: first an analytic variation of M for negative H with M being also negative; second a discontinuous jump from $-M(0)$ to $+M(0)$ for zero field; third a further analytic variation for positive field with M also positive. Translating this into the new language this implies the following. The isotherm of the lattice fluid below T_c consists of three pieces. The first piece, corresponding to negative field and magnetic moment, yields, because of (4.69), a density less than $1/2v$; the third, corresponding to positive field, yields a density larger than

$1/2v$; the discontinuity of the spontaneous magnetization yields a discontinuity in density at $\eta = 0$, that is at fixed pressure. This is the familiar condensation phenomenon. From (4.69) it follows that the densities d_1 and d_2 of the two phases in equilibrium obey the relation

$$d_1 + d_2 = 1/v. \qquad (4.72)$$

Above the temperature T_c the isotherm becomes a single analytic piece, like its analogue, the magnetization curve above the Curie point. The spontaneous magnetization at zero field, considered as a function of temperature, is thus imaged on the density of the lattice gas as a function of temperature when in equilibrium with its condensed phase; this density relation comes to a natural termination at the critical point T_c, which is the image of the Curie point.

This example teaches a valuable lesson concerning first and second order phase changes in general. The difference between them does not go as deep as the early theoretical studies indicated; a simple shift in the mode of observing the change can alter its character. Sublimation of a solid is to us a first order phase change, because we are used to observing it in an inert ambient of fixed pressure. If we were accustomed to evaporating a substance by heating it under its vapour pressure, then the process would appear to us as a second order phase change. Inversely, O. K. Rice[†] has shown that the two-dimensional Ising ferromagnet exhibited a second order phase change because it was examined theoretically at constant lattice spacing a. If the computations are made for constant pressure the phase change becomes first order whenever $dJ/da \neq 0$. If the coupling J is a function of a such that the derivative vanishes at some particular a, the phase change remains second order at this isolated point. This seems to be a general tendency: the second order change is the exception, the first order change the rule. Often, but not always, the former is a special point on a curve describing the latter.

OTHER SECOND ORDER PHASE CHANGES

It is evident from this analysis that the Ising model furnishes for the vast majority of co-operative phase changes a picture which is

† O. K. Rice, *Journ. Chem. Phys.* **22** (1954), 1535.

both qualitatively correct and internally consistent. A few second order phase changes remain to be explained, notably the transition of many metals to the superconducting state, and of liquid helium to the superfluid state. In both cases the ordered phase must be understood first before we can investigate seriously the nature of the transition. In both cases, too, we deal with a fluid which combines order and high fluidity. Solid state theory in general sees these two things as antagonistic. Present-day theory seeks to resolve this dilemma by looking for co-operative action in velocity space, but not in co-ordinate space. This would set these two co-operative phenomena apart from all the others if it should be finally confirmed.

PROBLEMS

1. The basic piece of information necessary to apply thermodynamics to ferromagnetism is that $-\mathbf{H}\cdot d\mathbf{M}$ is the work expended in producing in a sample a moment $\mathbf{M}$ with the help of a field $\mathbf{H}$. Introduce this into thermodynamics to establish the relation

$$C_H - C_M = -T\left(\frac{\partial M}{\partial T}\right)_H \left(\frac{\partial H}{\partial T}\right)_M.$$

Here C_H and C_M are the heat capacities of the sample at constant field and constant moment respectively. Use the result to derive (4.15) from (4.9) and (4.14).

2. Parelectric susceptibility arises from permanent electric dipoles capable of any orientation in space. Calculate the electric polarization as a function of temperature and field for an assembly of such identical independent dipoles, and point out its relationship, if any, to the magnetic formulas given.

3. Derive the specific heat formula analogous to (4.15) if L_∞ is the correct Langevin function. Comment on the value of the specific heat at $T=0$.

4. For zero magnetic field, $M=0$ is always a solution of (4.35). On what grounds can the solution be rejected below the Curie point?

5. Prove that the internal field theory predicts the change of entropy of a ferromagnet through the Curie point correctly; in other words, show that the value (4.17b) results for the integral (4.16b), even if computed from (4.32) and (4.36).

6. It was pointed out in the text that the simplest form of antiferromagnetism arises if the atoms can be assigned to two equivalent sublattices in such a way that only 2's are nearest neighbours of 1's and vice versa. Check the simple cubic, body-centred cubic, face-centred cubic and diamond structures, to find out which of these structures is suitable for antiferromagnetism from this point of view.

7. Show that in the internal field theory of antiferromagnetism the temperature rate of increase of the susceptibility immediately below the Curie point is double the rate of decrease immediately above.

8. Use the principles of Chapter 2 to derive all possible superstructure lines for β-brass.

9. Suppose we have an alloy with three nearest-neighbour interaction energies; v_{AA} for two neighbours of species A, v_{BB} for two neighbours of species B, and v_{AB} for two neighbours of opposite species. Show that the system is still equivalent to the Ising model with

$$J = \tfrac{1}{2}v_{AB} - \tfrac{1}{4}v_{AA} - \tfrac{1}{4}v_{BB}.$$

10. Prove the statement made in the text that the expression (4.46) is the internal energy for a linear Ising chain.

11. Prove that the expression (4.43) can have only one possible singular temperature value, and that this temperature is in fact given by formula (4.49).

12. Prove the correctness of equation (4.65) for the grand partition function Z from thermodynamics and from the expressions (4.63) and (4.64) for the derivatives of Z.

13. The reduction of the lattice gas to the Ising model showed that two states related to each other by the replacement of atoms by voids and voids by atoms have closely related properties in the lattice gas model. Show that the η's for the two states are equal and of opposite sign, and that $|\eta|$ is equal to $v\,\Delta p/kT$, where Δp is the difference between the pressures of the two states.

BIBLIOGRAPHY

GENERAL BACKGROUND

National Research Council. Committee on Solids. *Phase Transformations in Solids. Symposium held at Cornell University, Aug. 23–26, 1948.* (New York, Wiley, 1951.)

R. BECKER. *Theorie der Elektrizität*, vol. 2. (Leipzig, Teubner, 1933.)

J. H. VAN VLECK. *Theory of Electric and Magnetic Susceptibilities.* (Oxford, Clarendon Press, 1932.)

L. F. BATES. *Modern Magnetism*, 3rd ed. (Cambridge, University Press, 1951.)

FERROMAGNETISM

R. BECKER and W. DÖRING. *Ferromagnetismus.* (Berlin, Springer, 1939.)

E. C. STONER. Ferromagnetism. *Rep. Progr. Phys.* 11 (1946), 43–112; 13 (1950), 83–183.

C. KITTEL and J. K. GALT. Ferromagnetic domain theory. *Solid State Physics*, 3 (1956), 437–564.

K. H. STEWART. *Ferromagnetic Domains.* (Cambridge, University Press, 1954.)

ANTIFERROMAGNETISM

T. NAGAMIYA, K. YOSIDA and R. KUBO. Antiferromagnetism. *Advances in Physics*, **4** (1955), 1–112.

A. LIDIARD. Antiferromagnetism. *Rep. Progr. Phys.* **17** (1954), 201–44.

HIGH-LOW TRANSITIONS AND FERROELECTRICITY

R. B. SOSMAN. *Properties of Silica.* (Rutgers University Press, New Brunswick, 1959.)

P. W. FORSBERGH, JR. Piezoelectricity, electrostriction and ferroelectricity. *Handbuch der Phys.* **17** (1956), 264.

G. SHIRANE, F. JONA and R. PEPINKSY. Some aspects of ferroelectricity. *Proc. Inst. Radio Engrs, N.Y.* **43** (1955), 1738–93.

W. KÄNZIG. Ferroelectrics .and antiferroelectrics. *Solid State Physics*, **4** (1957), 1–197.

SUPERSTRUCTURE

F. C. NIX and W. SHOCKLEY. Order-disorder transformations in alloys. *Rev. Mod. Phys.* **10** (1938), 1–71.

T. MUTO and Y. TAKAGI. The theory of order-disorder transitions in alloys. *Solid State Physics*, **1** (1955), 193–282.

L. GUTTMAN. Order-disorder phenomena in metals. *Solid State Physics*, **3** (1956), 145–223.

ISING MODEL

G. F. NEWELL and E. W. MONTROLL. On the theory of the Ising model of ferromagnetism. *Rev. Mod. Phys.* **25** (1953), 353–89.

SPIN WAVES

G. HELLER and H. A. KRAMERS. *Proc. Amst. Acad. Sciences*, **37** (1934), 378.

M. J. KLEIN and R. S. SMITH. *Phys. Rev.* **80** (1950), 1111.

CHAPTER 5

ONE-ELECTRON THEORY OF
SOLIDS; METALS

Solids are one of the condensed states of matter. The atomic nuclei and the electrons form a compact whole with all parts tightly coupled. Physical theories treating such systems are very hard to develop, but it is not impossible to make progress, if formal theoretical procedures are coupled with physical insight. The most important insight in the present instance is the fact that the nuclei in a solid act almost like bodies at rest; this is evident from the geometrical studies of the first two chapters. When this feature is introduced into the general theory, it reduces to a dynamical theory of electrons in the stationary potential of the nuclei. As these considerations apply also to molecules, our present problem is seen to be closely linked to the problem of molecules. However, molecules are themselves an awkward theoretical problem, and thus the help we can get from that quarter is not great. Fortunately there are at least some solids which are simple enough to permit us to by-pass the molecular analogy, and to look to the case of atoms for inspiration.

An atom consists of a single atomic nucleus, essentially at rest, which is surrounded by a cloud of electrons. The field due to the nucleus is sufficiently strong to decouple the electrons from each other. This means that the electron-electron interaction simply modifies the nuclear field to a first approximation, and each electron can be assumed to move independently in a screened Coulomb field. The problem of atomic structure is thus reduced to solving several Schrödinger equations for one particle moving under the influence of a central force. This is a soluble problem of quantum mechanics which has easily recognizable qualitative features. The qualitative features account for Mendeleieff's periodic table of elements, and the quantitative treatment of the radial wave equation by the methods of Hartree or Fock leads to results which reproduce quantitatively many properties of atoms.

This precedent encourages us to expect some insight if we assume the electrons of a crystal uncoupled and make them move in a common crystalline potential field. The outstanding feature of this potential is of course that it has the symmetry of the crystal. Our previous experience would lead us to expect that among these symmetries the translational one is of greatest importance. Thus we would expect to get some useful information by solving the Schrödinger equation for a particle in a periodic potential. We shall see that the wave function for such a potential has certain features in common with the wave function for a free particle, notably the plane wave character, and the non-zero expectation value of the velocity.

WAVE MECHANICS OF ONE-DIMENSIONAL PERIODIC POTEN-TIALS

Following the precedent of Chapter 3 we shall elucidate first the nature of the wave functions for a potential periodic in one dimension. Let $V(x)$ be the potential in question and let d be its elementary period. We have then by assumption

$$V(x+d) = V(x). \tag{5.1}$$

The Schrödinger equation is

$$-\frac{\hbar^2}{2m}\frac{d^2\psi}{dx^2} + V(x)\,\psi = E\,\psi. \tag{5.2}$$

Denote by $u(x)$ and $v(x)$ two real linearly independent solutions of (5.2). We can use these to construct a number of others with the help of (5.1), such as $u(x+d)$ and $v(x+d)$. Since an ordinary differential equation of the type (5.2) has only two linearly independent solutions, these new solutions must be linear combinations of the old ones:

$$\left.\begin{aligned}
u(x+d) &= \alpha\,u(x) + \beta\,v(x), \\
v(x+d) &= \gamma\,u(x) + \delta\,v(x),
\end{aligned}\right\} \tag{5.3}$$

$$\alpha,\ \beta,\ \gamma,\ \delta \quad \text{real}. \tag{5.4}$$

Now from the two equations

$$-\frac{\hbar^2}{2m}\frac{d^2u}{dx^2} + V(x)\,u = E\,u,$$

$$-\frac{\hbar^2}{2m}\frac{d^2v}{dx^2} + V(x)\,v = E\,v,$$

we obtain by respective multiplication with v and u, and subtraction and integration, the so-called Green's identity, namely

$$\mathcal{W}(x) = v\frac{du}{dx} - u\frac{dv}{dx} = \text{constant}.$$

$\mathcal{W}$ is called the wronskian of the two solutions. Since it is a constant is must have the same value at $x+d$ and at x. However, from (5.3) we get

$$\mathcal{W}(x+d) = v(x+d)\frac{du(x+d)}{dx} - u(x+d)\frac{dv(x+d)}{dx}$$

$$= (\alpha\delta - \beta\gamma)\left\{v(x)\frac{du(x)}{dx} - u(x)\frac{dv(x)}{dx}\right\}$$

$$= (\alpha\delta - \beta\gamma)\mathcal{W}(x).$$

Hence the determinant of the coefficients of (5.3) must equal unity:

$$\alpha\delta - \beta\gamma = 1. \tag{5.5}$$

Suppose now we look for special solutions of (5.2) which are only multiplied by a factor upon displacement by d:

$$\psi(x+d) = \lambda\,\psi(x), \tag{5.6}$$

or $\qquad p\,u(x+d) + q\,v(x+d) = \lambda(p\,u(x) + q\,v(x)).$

Insertion of this into (5.3) yields the relations

$$\left.\begin{aligned}\alpha p + \gamma q &= \lambda p, \\ \beta p + \delta q &= \lambda q,\end{aligned}\right\} \tag{5.7}$$

which, upon elimination of p and q become

$$\begin{vmatrix} \alpha - \lambda & \gamma \\ \beta & \delta - \lambda \end{vmatrix} = 0. \tag{5.8}$$

Because of (5.4) and (5.5) this equation yields two roots which have a real sum and are reciprocals of each other. Hence there are three possibilities for these roots:

(a) The two roots are real and distinct. One is then absolutely greater, the other smaller than unity. Both solutions of (5.2) are divergent, one for large positive, the other for large negative x. According to the general criterion for wave functions these solutions are not acceptable, and the energy is not an eigenvalue.

(b) The two roots are conjugate complex and on the unit circle. The two roots are then conveniently written in the form

$$\lambda_+ = e^{2\pi i k d}, \qquad (5.9a)$$

$$\lambda_- = e^{-2\pi i k d}. \qquad (5.9\,b)$$

The wave functions are both allowed, and E is an eigenvalue with double degeneracy. In physics these solutions are generally called Bloch functions; we shall denote them by $b(x; k)$. The factor λ is called the Floquet factor of the solution after the original discoverer of this phenomenon.† Because of its suggestiveness, physicists prefer to specify the Bloch functions by their wave number k, as defined in (5.9); k has of course the usual indeterminacy of being defined only up to an additive multiple of $1/d$ (p. 53). We can utilize the wave number to write $b(x; k)$ in the form

$$b(x; k) = e^{2\pi i k x} u(x; k); \qquad (5.10)$$

$u(x; k)$ is then a periodic function of x with the period d.

(c) The root is double and equal to ± 1. The wave function is allowed, but the Floquet theorem which we just proved does not assure us that there are two such solutions.

It follows from (a), (b) and (c) that all possible values of λ are located in pairs such as Q and Q' on the real axis, or pairs such as P and P' on the unit circle, or at the points $+1$ or -1; this is shown in fig. 5.1. The topology in which they follow each other as the energy is changed is rather important in physical reasoning. In the first place, it is clear from continuity reasons alone that the solutions of type (a) and (b) occur in bands, separated by discrete energy states of type (c). The structure of the allowed bands of type (b) is mainly of interest in physics. We shall show now that in any one band λ cannot reverse the sense in which it varies along the unit circle. To prove this we must first check that $b(x; k)$ has no real root x_0; for if it did we could make it real at that point, together with its derivative, by multiplication with a complex factor; as a solution of a real equation with real initial conditions it would then be real altogether, and could not have a complex Floquet factor. With this preliminary information, we can study the nature

† G. Floquet, *Ann. sci. Éc. norm. sup. Paris*, **12** (1883), 47.

of the variation of the wave number k with the energy E by considering two points k and $k+\epsilon$ very close together; at either point, equation (5.2) is satisfied:

$$-\frac{\hbar^2}{2m}\frac{d^2 b(x;k)}{dx^2} + V(x)\,b(x;k) = E(k)\,b(x;k),$$

$$-\frac{\hbar^2}{2m}\frac{d^2 b(x;k+\epsilon)}{dx^2} + V(x)\,b(x;k+\epsilon) = E(k+\epsilon)\,b(x;k+\epsilon).$$

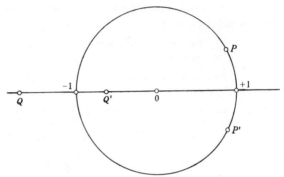

Fig. 5.1. Wave mechanics of a potential periodic in one dimension; location of the Floquet factor in the complex plane; P, P' and Q, Q' denote typical pairs of Floquet factors.

By forming and integrating Green's identity between them we get

$$b(x;k)\frac{db(x;k+\epsilon)}{dx} - b(x;k+\epsilon)\frac{db(x;k)}{dx}$$
$$= -\frac{2m}{\hbar^2}\{E(k+\epsilon) - E(k)\}\int^x b(x;k+\epsilon)\,b(x;k)\,dx. \quad (5.11)$$

If our notion is right that k cannot reverse when real, then k and E must advance together without changing their sense. We shall show this by proving that

$$\frac{b^2(x;k)}{b(x;k+\epsilon)},$$

which is, by construction, a function of wave vector $k-\epsilon$, solves the Schrödinger equation with an energy

$$2E(k) - E(k+\epsilon),$$

up to bounded correction terms which are quadratic in

$$E(k+\epsilon) - E(k).$$

Indeed we find

$$\left\{-\frac{\hbar^2}{2m}\frac{d^2}{dx^2}+V(x)\right\}\frac{b^2(x;k)}{b(x;k+\epsilon)} = \{2E(k)-E(k+\epsilon)\}\frac{b^2(x;k)}{b(x;k+\epsilon)}$$
$$-\frac{\hbar^2}{m}\left(\frac{1}{b(x;k+\epsilon)}\right)^3 [b(x;k+\epsilon)\,b'(x;k)-b(x;k)\,b'(x;k+\epsilon)]^2.$$

The square bracket can be substituted from (5.11) so that we finally get

$$\left\{-\frac{\hbar^2}{2m}\frac{d^2}{dx^2}+V(x)\right\}\frac{b^2(x;k)}{b(x;k+\epsilon)} = \{2E(k)-E(k+\epsilon)\}\frac{b^2(x;k)}{b(x;k+\epsilon)}$$
$$-\frac{4m}{\hbar^2}\left(\frac{1}{b(x;k+\epsilon)}\right)^3 \{E(k+\epsilon)-E(k)\}^2 \left[\int^x b(x;k+\epsilon)\,b(x;k)\,dx\right]^2.$$
$$(5.12)$$

The first line is the desired relation which implies quasilinear advance of k with E. The second line is quadratic in the energy difference as predicted; the cube of the reciprocal wave function is bounded, since the wave function was shown to have no root. The boundedness of the integral is the most delicate part. It follows from (5.9) that the integrand is multiplied by $e^{2\pi i(2k+\epsilon)d}$ as x is increased by d. If this multiplier is 1 in the limit $\epsilon = 0$, the integral is divergent for large x; otherwise it is oscillatory and bounded. We see from (5.9) that the possible divergencies correspond to $\lambda = \pm 1$. Hence we see that λ can reverse direction only at ± 1 if at all. Each point on the half circle above the real axis occurs, therefore, just once, and with it is coupled, because of (5.9), the complex conjugate value $1/\lambda$ below the axis. In terms of k it means that positive and negative k's are associated with the same energy, and the variation of energy with k is monotone in each band as k varies from 0 to $1/2d$. This is the same situation as we had for lattice vibrations, as seen from the figs. 3.2 and 3.3. Every band is thus a complete band in the reduced Brillouin zone and we may conveniently label these bands by a supplementary index q in the sense of increasing energy:

$$q = 1, 2, 3, 4, \ldots \quad \text{as } E \text{ increases.}$$

A complete description of wave functions and eigenvalues for the equation (5.2) thus takes the form

$$\psi = b_q(x;k) \qquad (5.13)$$

and
$$E = W_q(k). \tag{5.14}$$

For every q, these functions are defined for all k. They obey the symmetry relations

$$b_q(x+d; k) = e^{2\pi i k d} b_q(x; k), \tag{5.15}$$

$$b_q\left(x; k+\frac{1}{d}\right) = b_q(x; k), \tag{5.16}$$

$$W_q\left(k+\frac{1}{d}\right) = W_q(k), \tag{5.17}$$

$$b_q(x; -k) = b_q^*(x; k), \tag{5.18}$$

$$W_q(-k) = W_q(k), \tag{5.19}$$

which are evident from the preceding discussion. The appearance of the conjugate complex in (5.18) arises from the interrelation of the two roots exhibited in (5.9), together with the reality of the basic equation (5.2).

It is worthwhile to supplement this abstract analysis by simple examples. The simplest analytical example is

$$V = V_0 \cos\frac{2\pi x}{d}, \tag{5.20}$$

to be substituted into (5.2). The resultant equation is the *Mathieu differential equation*. Mathieu functions do indeed exhibit all the properties (5.15) to (5.19). However, working with them is not easy, and as a consequence, the sinusoidal potential does not have the value for crystal physics which the harmonic oscillator or the hydrogen atom have in other branches of physics. We shall take up here instead a non-analytic example, the so-called *Kronig–Penney model*, which is easy at least in its early stages. The potential of the Kronig–Penney model is shown in fig. 5.2. Potential free regions are seen to be separated by square potential wells of height V and thickness t, located at regular intervals of period d. An even simpler subcase to which discussion will be limited is the so-called *δ-well model* in which V is very large and t very small, but their product retains a finite value. The Schrödinger equation for the δ-well model is equation (5.2), which, outside the wells, reduces simply to

$$\frac{d^2\psi}{dx^2} + \alpha^2\psi = 0, \tag{5.21a}$$

or
$$\frac{d^2\psi}{dx^2} - \beta^2\psi = 0, \qquad (5.21b)$$

according to whether the energy is positive or negative. Here the parameters α and β are defined as intrinsically positive quantities equal to

$$\alpha = \left(\frac{2mE}{\hbar^2}\right)^{\frac{1}{2}}, \qquad (5.22a)$$

and
$$\beta = \left(\frac{2m(-E)}{\hbar^2}\right)^{\frac{1}{2}}. \qquad (5.22b)$$

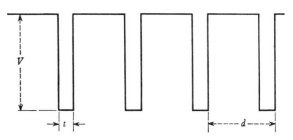

Fig. 5.2. The δ-well model, a soluble case of a one-dimensional periodic potential.

Within the wells $-V$ may be presumed so large that the term in E in (5.2) can be neglected. Moreover, infinite V implies infinite curvature in the wave function, that is a corner. Thus the wave functions take the scalloped aspect shown in fig. 5.3, and the only information required from inside the well is their change in slope which results from it. This may be simply secured by integration of (5.2) across the well. We find

$$\left(\frac{d\psi}{dx}\right)_{\text{right}} - \left(\frac{d\psi}{dx}\right)_{\text{left}} = -2\eta\psi, \qquad (5.23)$$

where
$$\eta = \frac{mVt}{\hbar^2}. \qquad (5.24)$$

By assumption, this quantity η has a fixed value in the limit of infinite V.

To solve the problem formulated through the equations (5.21) and (5.23), locate one of the barriers at the origin $x = 0$; another one is then located at $x = d$. We can restrict consideration to this open interval, excluding all singularities, by combining (5.15) and (5.23) in the form

$$\psi(0) - e^{-2\pi ikd}\psi(d) = 0, \tag{5.25}$$

$$\left(\frac{d\psi}{dx}\right)_{x=0} - e^{-2\pi ikd}\left(\frac{d\psi}{dx}\right)_{x=d} = -2\eta\psi(0). \tag{5.26}$$

We satisfy (5.21) by writing ψ as a sum of two sines vanishing at either end. (5.25) then determines the coefficients, except for normalization. We get

$$\psi(x) = \sin\alpha(d-x) + e^{2\pi ikd}\sin\alpha x, \tag{5.27a}$$

or $\qquad \psi(x) = \sinh\beta(d-x) + e^{2\pi ikd}\sinh\beta x. \tag{5.27b}$

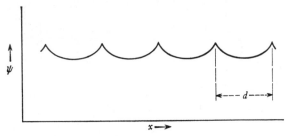

Fig. 5.3. Scalloped aspect of the wave function for the δ-well model.

Substitution of (5.27) into (5.26) yields then the final eigenvalue condition

$$\cos 2\pi kd = \cos\alpha d - (\eta/\alpha)\sin\alpha d, \tag{5.28a}$$

or $\qquad \cos 2\pi kd = \cosh\beta d - (\eta/\beta)\sinh\beta d. \tag{5.28b}$

Equation (5.28) may be analysed by assuming some fixed value for ηd and plotting the right-hand side as function of αd and βd respectively. This is done in fig. 5.4 for $\eta d = 1$. We see that we get an oscillatory curve with an amplitude of oscillation which is larger than 1. Equation (5.28) picks from this curve those values of α and β for which the function is less than 1 and for which a real value of k can thus be found. To get the energy we use (5.22); it is seen that to plot energy instead of αd and βd on abscissa would simply distort the abscissa scale, but would leave qualitative conclusions unaffected. Examination of fig. 5.4 shows that for low energy the energy band is relatively narrow and the forbidden gap wide. As the energy increases the allowed bands become wider and the forbidden gaps narrower. These forbidden gaps never disappear completely, no matter how much the energy increases. It is also seen that in each of the allowed bands kd takes up exactly once any particular value between $-\frac{1}{2}$ and $+\frac{1}{2}$, as was pointed out in the general analysis preceding this example.

Having analysed the δ-well model to check the structural features of the ψ-function for a periodic potential, we can point the analysis

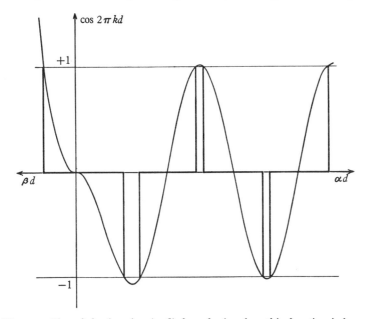

Fig. 5.4. Plot of the function (5.28) for $\eta d = 1$; when this function is larger than 1 or smaller than -1 the state is forbidden; when it is between -1 and $+1$ its arccos determines the wave number.

more specifically toward solids. An isolated δ-well of fixed η has a single negative energy state which results from the simultaneous solution of (5.23) at the origin and of (5.21b) everywhere else. The wave function is

$$\psi = e^{-\eta|x|}, \qquad (5.29)$$

and the energy from (5.22b)

$$E = -\frac{\hbar^2 \eta^2}{2m}. \qquad (5.30)$$

The δ-well can therefore serve as a model of an atom with a bound state, and the periodic system of δ-wells of fig. 5.2 as a model of a crystalline collection of such atoms. In considering such crystals, theoretical physicists have found it useful to consider the electronic states as a function of the lattice parameter. When the lattice parameter is very large the atomic levels are essentially undisturbed, and if thereupon the interatomic distance is decreased the levels widen into bands. We can duplicate this kind of investigation on the bound state of the δ-well model, treating η as fixed and studying E for the band edges as functions of d; by our general theory these edges are at $k = 0$ and $k = 1/2d$ respectively. Substitution of this into (5.28) yields for the β_L of the lower edge

$$\coth \tfrac{1}{2}\beta_L d = \beta_L/\eta \qquad (5.31)$$

and for β_U or α_U of the upper edge

$$\tanh \tfrac{1}{2}\beta_U d = \beta_U/\eta, \qquad (5.32a)$$

$$\tan \tfrac{1}{2}\alpha_U d = \alpha_U/\eta. \qquad (5.32b)$$

To get the energies, the α's and β's resulting from (5.31) and (5.32) are to be substituted into (5.22). The resultant E versus d curves are shown in fig. 5.5. We see a characteristic widening out of the band with decreasing interatomic distance. The example suggests that the process of reaching the actual crystal through decreasing lattice constants can be used as an adiabatic device to label the crystalline bands in terms of the bound atomic levels from which they arise in this process.†

The relationship of bands to bound states is only one aspect of their two-sided nature. They can also be related to the energy continuum of a free particle. To bring this out we consider the value of d fixed in the equations (5.28) and let η tend to zero. There is no advantage in doing this with a particular model such as the δ-well model, since the procedure is perfectly general. Let us therefore replace the model by the generalized expression

$$V(x) = \eta \sum_{\nu=-\infty}^{+\infty} v_\nu \, e^{2\pi i \nu x/d}, \qquad (5.33)$$

---

† The device does not always work. Already in our example it fails for all higher bands which arise from the continuum. In three dimensions, the order of the levels in energy can change as the lattice parameter is varied; the device may then fail even for low lying bands.

which is to be considered a perturbation acting on free-particle wave functions; this means that we should treat the strength parameter η in (5.33) as small. The unperturbed wave functions of the problem are thus the plane waves

$$\psi(x; p) = e^{ipx/\hbar}, \qquad (5.34)$$

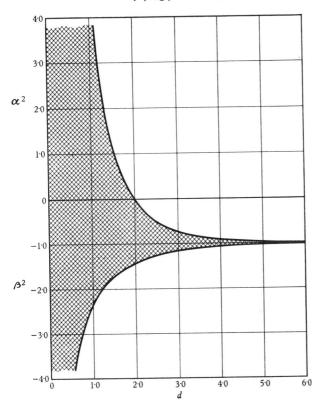

Fig. 5.5. Plot of the energy of the lowest band in the δ-well model as function of the lattice spacing. The bound discrete state of an isolated δ-well merges into the lowest band as the other δ-wells approach.

with the energies $\qquad E(p) = p^2/2m.$ $\qquad (5.35)$

In this limiting case we have the relation

$$p = \hbar k, \qquad (5.36)$$

relating the wave number to the momentum p. A great deal of the qualitative interpretation of wave functions obeying (5.15)–(5.19) is associated with the relation (5.36) and the sense in which it is true in the presence of a periodic potential. Let us, therefore, introduce the potential (5.33) to see how it affects the free-particle wave functions. The matrix elements which V generates among the plane waves are

$$\langle p' | V(x) | p \rangle = \frac{\eta}{Ld} \sum_{\nu = -\infty}^{+\infty} \int_0^{Ld} e^{i(p-p')x/\hbar} v_\nu e^{2\pi i \nu x/d} \, dx. \qquad (5.37)$$

Here Ld is a Born–von Karman period assumed for convenience; it implies that p and p' are integral multiples of h/Ld. All integrals then vanish which do not satisfy the condition

$$p' - p = \nu(h/d), \qquad (5.38a)$$

or, because of (5.36) $k' - k = \nu(1/d).$ \qquad (5.38b)

This means, that after mixing of wave functions has taken place, wave numbers differing by an integral multiple of $1/d$ are not distinguishable. This is an indeterminacy familiar from the theory of lattice vibrations (equations (3.13) and (3.17)); in three dimensions it widens out into the Bragg–Laue condition (2.12) and the reducibility of all wave vectors to the first Brillouin zone, as discussed in Chapters 2 and 3. Once this is realized, then (5.38) imposes no further mixing. In other words the reduced wave vector is a constant of the motion; this is obvious from the group theoretical discussion of Chapter 3. An interesting corollary to this is the fact that all expressions of the form

$$e^{2\pi i n k d} = e^{2\pi i n p d/h}, \qquad (5.39)$$

with arbitrary integer n, have the same value for all of the mixed wave functions. The expressions (5.39) are called the *crystal translation operators* and are *exact* constants of the motion. They provide the sense in which (5.36) is true and in which momentum is conserved.

Shifting our attention from the momentum to the energy we see that, in general, the potential contributes a simple first order shift to each level; the amount is ηv_0, the value of the matrix element (5.37) in the diagonal case. Exceptional points exist, however; those momenta p which have matrix elements with the state $-p$ yield a degenerate perturbation problem. These exceptional states are the ones for which a Bragg reflection also conserves energy; it is natural, therefore, to expect that the perturbation of these states will be particularly severe. Let n be an integer; the momentum

$$p_0 = \tfrac{1}{2}nh/d \tag{5.40}$$

is then such an exceptional momentum, which is coupled to its negative, as is seen from (5.38); let $\pm p_0 + \delta p$ be a momentum in the immediate vicinity of these two special states. The coupling between these two is made by the Fourier component v_n as is seen from (5.37). The result is a two-by-two energy matrix which equals

$$\mathscr{H} = \begin{pmatrix} \dfrac{1}{2m}(p_0+\delta p)^2 + \eta v_0 & \eta v_n \\ \eta v_n^* & \dfrac{1}{2m}(-p_0+\delta p)^2 + \eta v_0 \end{pmatrix}. \tag{5.41}$$

At the position $\delta p = 0$, the energy level is split according to the scheme

$$E = \frac{1}{2m}p_0^2 + \eta v_0 \pm \eta\,|v_n|. \tag{5.42}$$

The split destroys the energy continuum. This is seen most easily if the difference between the two energies is calculated as a function of δp; we find

$$(\Delta E)^2 = 4\left[\eta^2\,|v_n|^2 + \frac{1}{m^2}\,p_0^2(\delta p)^2\right]. \tag{5.43}$$

This is the equation of a hyperbola when considered in a $\Delta E - \delta p$ plane. Its two branches represent raised and lowered energy; they are separated for all δp, with the minimum separation (5.42) at $\delta p = 0$. The free-energy parabola is thus cut up into separated segments which along the k axis have length $1/2d$. This is shown in fig. 5.6 for the δ-well model. Each piece is monotone in E as required by the theory. Conventional bands are constructed out of these pieces by combining sections of equal energy range. The result is the band picture of fig. 5.7, for which all relevant information is within the first Brillouin zone, that is between the values $k = -1/2d$ and $k = +1/2d$. This type of periodic picture is familiar to us from the figs. 3.2 and 3.3. The new element here is that the number of bands is infinite.

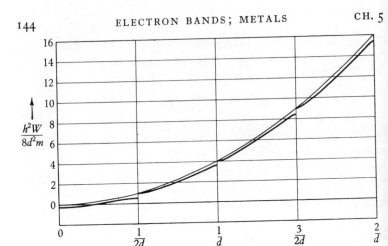

Fig. 5.6. Plot of energy versus momentum for a free particle and for the δ-well model at $\eta d = 1$. The δ-wells cut up the free energy parabola into segments whose length is determined by the Bragg conditions.

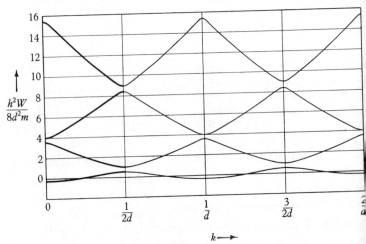

Fig. 5.7. Transformation of the curve 5.6 into the periodic relationship between energy and wave number which is typical for periodic potentials.

WAVE MECHANICS OF THREE-DIMENSIONAL PERIODIC POTENTIALS

It is seen from the preceding that the band concept for periodic potentials in one dimension is capable of relatively precise formulation. The case of three dimensions is not quite so favourable in certain respects, but the essentials can indeed be generalized. To begin with, the plane wave property for ψ functions, holds also for the solutions of the three-dimensional Schrödinger equation

$$-\frac{\hbar^2}{2m}\nabla^2\psi + V(\mathbf{x})\psi = E\psi, \qquad (5.44)$$

if the potential $V(\mathbf{x})$ obeys the periodicity condition

$$V(\mathbf{x}+\mathbf{r}) = V(\mathbf{x}) \qquad (5.45)$$

for any $\mathbf{r}$ given by (1.2). The reason is the group theoretical one developed in Chapter 3. Indeed, the set of solutions of (5.44) form, for each energy value E, a set of quantities X_i to which the theorem of p. 62 is applicable. Their spatial dependence has therefore the form (3.32), which is more simply expressed by introduction of the wave vector (3.37). We write the wave functions and energies in the form

$$\psi = b_q(\mathbf{x};\mathbf{k}), \qquad (5.46)$$

$$E = W_q(\mathbf{k}). \qquad (5.47)$$

They obey the symmetry relations

$$b_q(\mathbf{x}+\mathbf{r};\mathbf{k}) = e^{2\pi i \mathbf{k}\cdot\mathbf{r}}b_q(\mathbf{x};\mathbf{k}), \qquad (5.48)$$

$$b_q(\mathbf{x};\mathbf{k}+\mathbf{r}^*) = b_q(\mathbf{x};\mathbf{k}), \qquad (5.49)$$

$$W_q(\mathbf{k}+\mathbf{r}^*) = W_q(\mathbf{k}), \qquad (5.50)$$

$$b_q(\mathbf{x};-\mathbf{k}) = b_q^*(\mathbf{x};\mathbf{k}), \qquad (5.51)$$

$$W_q(-\mathbf{k}) = W_q(\mathbf{k}). \qquad (5.52)$$

Although these relations look generally reasonable, in the sense that the transformed functions are also admissible solutions, there is some comment needed concerning the topology of these solutions in k-space. In the way the formulas are written this topology is hidden by the assignment of a 'band index' q which is presumed to complement the reduced wave vector $\mathbf{k}$ in specifying the quantum state. At a point of no degeneracy this assignment can be made by the use of the energy: we specify q by the relations

$$q = 1, 2, 3, 4, \ldots \qquad (5.53a)$$

and

$$q' > q, \quad \text{if} \quad W_{q'}(\mathbf{k}) > W_q(\mathbf{k}). \qquad (5.53b)$$

This prescription does, of course, break down at degeneracy points. In mitigation of this defect it may be said that these degeneracy points (that is points having two wave functions of equal $\mathbf{k}$ and W) form at most lines in $\mathbf{k}$-space, and never surfaces, and that the general line in this space will not meet any such points.† Therefore, along the general line the prescription (5.53) will coincide with the result of analytical continuation. This makes the choice intrinsically reasonable. Nevertheless, there remains a set of paths of measure zero which pass through a degeneracy point, and for which this identity of the two assignments is not verified. It is not worth while to extend the definition (5.53) so as to include such points, because the choice of wave functions ceases to be unique anyway. What is worth while, however, is to understand equation (5.47) a little better in such regions. The starting point is the observation that the determination of the eigenvalues is equivalent with the diagonalization of a matrix. In the neighbourhood of an m-fold degeneracy this means that we must diagonalize a matrix of rank m whose elements may be assumed analytic functions of $\mathbf{k}$. Before such a diagonalization is accomplished (5.46) and (5.47) are conveniently replaced by the more general situation

$$\mathscr{H}\beta_q^\mu(\mathbf{x};\mathbf{k}) = \sum_{\nu=1}^{m} w_q^{\mu\nu}(\mathbf{k})\,\beta_q^\nu(\mathbf{x};\mathbf{k}), \qquad (5.54$$

where the functions $\beta_q^\nu(\mathbf{x};\mathbf{k})$ are intermediate Bloch type function which are not fully diagonalized, and the $w_q^{\nu\mu}$ are coefficients having the property that at the degeneracy point (or line) $\mathbf{k} = \mathbf{k}_0$ all of diagonal coefficients vanish and the diagonal ones are equal (hierarchy of degeneracy for special points on a degenerate line i of course also possible).‡ In terms of these basis functions β_q^ν, th eigenvalues are then given by

$$\begin{vmatrix} w_q^{11} - W_q & w_q^{12} & w_q^{13} & \dots & w_q^{1m} \\ w_q^{21} & w_q^{22} - W_q & w_q^{23} & \dots & w_q^{2m} \\ \dots & & & & \\ \dots & \dots & \dots & \dots & w_q^{mm} - W_q \end{vmatrix} = 0. \qquad (5.55$$

W appears now as a root of an algebraic equation of degree m; it

† Exceptions to this rule are known, the most notable being the close-pack hexagonal structure. See C. Herring, *Phys. Rev.* **52** (1937), 361.

‡ W. Shockley, *Phys. Rev.* **78** (1950), 173.

known that such roots are generally not analytic functions (that is, developable in power series) of the coefficients or of any parameter contained in them if they are examined in the neighbourhood of a degeneracy point. On the other hand, the actual behaviour of the energy as a function of the wave vector is reduced by the relation (5.55) to an algebraic problem. Once this problem is solved the true Bloch functions

$$b_q(\mathbf{x}; \mathbf{k}) = \sum_{\nu=1}^{m} p_\nu \beta_q^\nu(\mathbf{x}; \mathbf{k}) \tag{5.56}$$

are then also obtainable by well-known methods; there are m such functions b_q, one corresponding to each root.

From this digression about degeneracy it is evident that, while there are certain difficulties of computation and interpretation for the equations (5.46)–(5.52) when they are studied near degeneracy points or lines, they are not such as to render the equations invalid. In particular, the symmetry relation (5.51)–(5.52) becomes a generally valid generalization of (5.17)–(5.18). This is a degeneracy in energy of two states with different $\mathbf{k}$ which is not required by crystalline symmetry. It is known as the *time inversion symmetry* inherent in the Schrödinger equation, and is the analogue for matter waves of the theorem in optics on the reversibility of the light path.

A second property which these wave functions share with free particle wave functions is that they describe a state of definite mean velocity. This velocity is obtained as follows. If we combine (5.46) and (5.47) to get a wave function with time dependence we get with the abbreviation (3.12)

$$\psi(\mathbf{x}, t) = u_q(\mathbf{x}; \mathbf{k}) \operatorname{ixp} \left[\mathbf{k} \cdot \mathbf{x} - \frac{W_q(\mathbf{k}) t}{h} \right]. \tag{5.57}$$

Here $u_q(\mathbf{x}; \mathbf{k})$ is a function similar to the one defined in (5.10); it has the period of the crystal. Let us make a wave packet from such waves which has a peak around a definite value $\mathbf{k}_0$ and which is sufficiently concentrated about it so that $W_q(\mathbf{k})$ can be replaced by its Taylor expansion to linear terms in $\mathbf{k}$:

$$\psi(\mathbf{x}, t) = \operatorname{ixp} \left[\mathbf{k}_0 \cdot \mathbf{x} - \frac{W_q(\mathbf{k}_0) t}{h} \right]$$
$$\times \iiint d\mathbf{k}\, u_q(\mathbf{x}; \mathbf{k})\, \phi(\mathbf{k}) \operatorname{ixp} \left[(\mathbf{k} - \mathbf{k}_0) \cdot \left(\mathbf{x} - \frac{t}{h} \frac{\partial W_q}{\partial \mathbf{k}_0} \right) \right]. \tag{5.58}$$

Here $\phi(\mathbf{k})$ is the distribution function which is assumed steeply peaked about $\mathbf{k}_0$. Clearly, the expression outside the integral is a phase factor which vanishes when we form $\psi^*\psi$. Apart from a modulation within the primitive cell arising from the periodic function $u_q(\mathbf{x};\mathbf{k})$ this quantity $\psi^*\psi$ will remain constant along a trajectory for which

$$\mathbf{x} = \frac{t}{h}\frac{\partial W_q}{\partial \mathbf{k}_0}.$$

This yields the velocity of the packet as

$$\mathbf{v} = \frac{1}{h}\frac{\partial W_q(\mathbf{k})}{\partial \mathbf{k}}, \qquad (5.59)$$

which is the desired formula. Formula (5.59) retains its simple form even if $\mathbf{v}$ and $\mathbf{k}$ are expressed in the reciprocal affine co-ordinates described by the equations (2.9). It enhances still further the identification suggested by equation (5.36). For if we interpret $h\mathbf{k}$ as a momentum and $W_q(\mathbf{p}/h)$ as a hamiltonian, then (5.59) is one-half of Hamilton's equations; the other half yields the constancy of $\mathbf{k}$ and is also verified. Further resemblances to the free particle situation arise if one considers the relation (5.47) near an energy minimum; if the minimum is not complicated by degeneracies of bands then the energy is developable in a power series in $\mathbf{k}$ with missing linear terms:

$$W_q(\mathbf{k}) = W_q(\mathbf{k}_0) + \tfrac{1}{2}\sum_{ij}\alpha_{ij}(k_i-k_{0i})(k_j-k_{0j})+\dots. \qquad (5.60)$$

The coefficients α_{ij} form something like a reciprocal mass tensor. This tensor, in turn, might even degenerate into a simple mass under special conditions. This specialization arises in particular for symmetry reasons, when the crystal is cubic and the energy minimum is located at $\mathbf{k}_0 = \mathbf{0}$:

$$W_q(\mathbf{k}) = W_q(\mathbf{0}) + \frac{h^2}{2m_q^*}k^2 + \dots. \qquad (5.61)$$

m_q^* is called the *effective mass* of the electrons at the bottom of the band q. Even if (5.61) is not quite true in a particular case its study still provides qualitative understanding.

FERMI STATISTICS AND ELECTRON DEGENERACY

No treatment of crystalline electrons as approximately free particles has any chance of success if the particles are treated as free in the sense of the kinetic theory of gases. Drude[†] tried to make the identification this way, as soon as he realized that electrons were the carriers of the electric current in metals. After some early successes in the field of transport properties, difficulties arose in the theory over the specific heat contribution arising from the conduction electrons. The electronic specific heat is much smaller, even for valence electrons, than the value $\frac{3}{2}R$ predicted by Drude's theory. Sommerfeld[‡] first showed that the free electron theory of metals will only make qualitative sense if it is coupled with the proper statistical considerations. The essential element in these considerations is the Pauli exclusion principle for electrons. The importance of this principle in producing the shell structure of the atom is well known. It is, therefore, not too surprising that it also plays a vital role in the energy bands which replace some of these atomic states in a crystalline aggregate.

The result of introducing the Pauli exclusion principle into statistical mechanics is the *Fermi–Dirac distribution* law. It reads[§]

$$n(\epsilon) = \frac{g(\epsilon)}{e^{(\epsilon-\eta)/kT} + 1}. \tag{5.62}$$

Here $n(\epsilon)$ is the number of electrons per unit energy range at the energy ϵ, $g(\epsilon)$ the number of states per unit energy range, and η the *Fermi energy*: a temperature-dependent parameter determined by the side condition

$$\int_{-\infty}^{+\infty} n(\epsilon)\, d\epsilon = N, \tag{5.63}$$

where N is the total number of electrons. The law contains the older Maxwell–Boltzmann law in two limiting forms. If

$$\epsilon - \eta \gg kT, \tag{5.64a}$$

then the 1 in equation (5.62) can be neglected, and a Boltzmann relation

$$n(\epsilon) \approx g(\epsilon)\, e^{-(\epsilon-\eta)/kT} \tag{5.64b}$$

[†] P. Drude, *Ann. Phys., Lpz.*, **1** (1900), 566.
[‡] A. Sommerfeld, *Z. Phys.* **47** (1928), 1.
[§] The law will not be derived here as it is much more appropriately treated in statistical mechanics. The reader is referred to treatises on that subject.

results. If, on the other hand, we have

$$\eta - \epsilon \gg 0, \qquad (5.65)$$

then the exponential in (5.62) is small and the limiting law

$$g(\epsilon) - n(\epsilon) \approx g(\epsilon)\, e^{(\epsilon - \eta)/kT} \qquad (5.65c)$$

results from an expansion in powers of that small exponential. To a first approximation $n(\epsilon)$ and $g(\epsilon)$ are equal in this case, that is, all states are filled. If we are willing, however, to treat the unoccupied states as if they were particles then they will have a Boltzmann

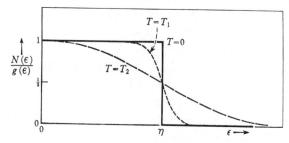

Fig. 5.8. Fermi statistics; plot of occupation probability versus energy; the curve has sharp corners at absolute zero and becomes more rounded at higher temperatures.

distribution with the energy turned upside down; this is the statistical part of the basis for the concept of *defect electrons* or *holes* which will be so prominent in semiconductor theory. In between the two limiting ranges there is a transition region

$$|\epsilon - \eta| \lesssim kT. \qquad (5.66)$$

In the middle of it, at $\epsilon = \eta$ the probability of occupation is always $\frac{1}{2}$. It is to be observed that the transition becomes more abrupt as the absolute zero is approached. Fig. 5.8 shows $N(\epsilon)/g(\epsilon)$ as a function of ϵ if T is varied; the Fermi energy η is taken as fixed. In reality, the Fermi energy drifts slowly with temperature. Its determining relation (5.63) is such as to make it drop with rising temperature, since $g(\epsilon)$ is a generally rising function of ϵ.

As a consequence of the band picture and the Fermi–Dirac distribution law we get a ready classification of solids into insulators and conductors. If, for a given solid, the band structure and the number of electrons per primitive cell is such that all energy bands are filled and an energy gap separates the topmost electrons from the lowest empty band, then the solid is an insulator. The reason for this is that the current carried by a full band is always zero, as one verifies by integrating (5.59) over a primitive cell of the reciprocal lattice, and making use of the periodicity of $W(\mathbf{k})$ in that lattice. Excitation of electrons out of these full bands cannot be achieved in infinitesimal steps since the gap separation has to be overcome. On the same picture, conductors will be those crystals in which, for one reason or another, some band is only partially filled. It is true that such an unfilled band still carries no total current in equilibrium, because it follows from (5.52) and (5.59) that the contributions to the current from individual electrons cancel in pairs; but disturbance from this equilibrium requires now only an infinitesimal expense of energy, which means that the kinetic kinds of considerations used for transport phenomena in gases are applicable to electrons in an unfilled band. The reasons for having partially filled bands are essentially twofold. Crystals with an odd number of electrons per cell have usually an unfilled band because the electron spin allows the placing of two electrons into each orbital; the alkali metals are outstanding examples of this situation. Second, even if the number of electrons per cell is even, unfilled bands can arise through the overlapping of bands. A well-analysed case of this kind arises for the iron group metals, for which the $3d$ band and $4s$ band are as shown in fig. 5.9 and neither band is filled until the element zinc is reached. Partially filled bands arise also for elements such as the alkaline earths (zinc in the case of fig. 5.9) which, at first thought, should have a filled s-band. In this case, the next higher band (usually a p band) overlaps slightly on the high energy side. This is confirmed by the study of these metals in X-ray emission where these bands can be observed experimentally.† Fig. 5.10 shows the empirical structure of the filled parts of the $3s$ and $3p$ bands so obtained for aluminium.

The electrons of a partially filled metallic band are in a condition far removed from that of an ideal maxwellian gas, not so much because of modifications in their mechanics, but mainly because their statistics are completely different. Near the absolute

† H. W. B. Skinner, *Phil. Trans. Roy. Soc.* **239** (1940), 95.

zero of temperature this is obvious, because the partial filling of states to a given level bears no resemblance to Boltzmann statistics. As the temperature rises the Fermi level drops, and the question whether Boltzmann statistics are applicable becomes a quantitative one; if Boltzmann statistics are not applicable the electrons are

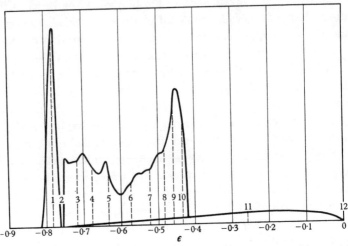

Fig. 5.9. Density of states in energy versus energy for the transition metals. The high curves arise from the $3d$ shell and the low one from the $4s$ shell; twelve electrons can be placed into the two combined shells (after Slater).

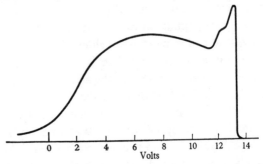

Fig. 5.10. Empirical shape of the filled parts of the $3s$ and $3p$ bands in aluminium, as observed in X-ray emission (after Skinner).

called *degenerate*. We shall now examine whether the electrons in the conduction band of a metal are degenerate at a given temperature. To simplify, we shall treat these electrons as a gas of effective mass m^*.

By a general principle of statistics, the number of quantum states available to an electron in a portion of phase space is the volume of that portion divided by h^3, or rather twice that amount if the spin is considered

$$\mathcal{N} = \frac{2}{h^3} \iiint\!\!\!\iiint dp_x\, dp_y\, dp_z\, dx\, dy\, dz. \tag{5.67}$$

To get the $g(\epsilon)$ employed in (5.62), which is appropriate for this case, (5.67) must be integrated over the space between two hyper-

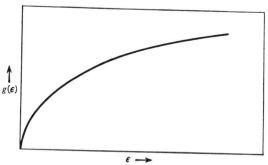

Fig. 5.11. Density of states in energy versus energy for quasi-free electrons.

surfaces, one of energy ϵ, the other of energy $\epsilon + d\epsilon$, as explained in the derivation of (3.39). This elementary calculation yields

$$g(\epsilon) = 0 \qquad \text{for} \quad \epsilon < 0, \tag{5.68a}$$

$$g(\epsilon) = \frac{2^{\frac{7}{2}}\pi m^{*\frac{3}{2}}\Omega}{h^3}\sqrt{\epsilon} \quad \text{for} \quad \epsilon > 0, \tag{5.68b}$$

where Ω is the metal volume. The curve is a parabola; it is exhibited in fig. 5.11. At the absolute zero, filling occurs up to the Fermi level $\eta(0)$ which is found by use of the version of (5.63) valid at absolute zero:

$$N = \int_{-\infty}^{\eta(0)} g(\epsilon)\, d\epsilon. \tag{5.69}$$

One finds from this, for the case that (5.68) is valid:

$$N = \tfrac{2}{3} g\{\eta(0)\}\, \eta(0), \tag{5.70}$$

which when solved with respect to $\eta(0)$, yields

$$\eta(0) = \frac{3^{\frac{2}{3}}h^2}{8\pi^{\frac{2}{3}}m^*}\left(\frac{N}{\Omega}\right)^{\frac{2}{3}}. \tag{5.71}$$

In order to verify if the degeneracy has disappeared at the temperature T we must find out if η has slipped out of the band under these conditions. To find a criterion for this we calculate the limiting temperature and density at which η is just at $\epsilon = 0$. For densities which are lower, or temperatures which are higher, than these limits the gas is no longer degenerate. We find the limiting density N_L by carrying out the integration (5.63) for $\eta = 0$:

$$N_L = \int_{-\infty}^{\infty} \frac{g(\epsilon)\,d\epsilon}{\exp(\epsilon/kT_L)+1}, \qquad (5.72)$$

which can also be written as

$$N_L = \int_{-\infty}^{+\infty} g(\epsilon)\,d\epsilon\,[e^{-\epsilon/kT_L} - e^{-2\epsilon/kT_L} + e^{-3\epsilon/kT_L} - \ldots].$$

Substitution of (5.68) for $g(\epsilon)$ yields then

$$N_L = \tfrac{1}{2}\pi^{\frac{1}{2}} g(kT_L)\, kT_L \sum_{\nu=1}^{\infty} (-)^{\nu}\left(\frac{1}{\nu}\right)^{\frac{3}{2}}$$

$$= 0\cdot678\, g(kT_L)\, kT_L.$$

Calling D_L the limiting density and substituting g from (5.68) we arrive thus at the numerical relation

$$D_L = 3\cdot85 \times 10^{15}\,(m^*/m)^{\frac{3}{2}}\, T_L^{\frac{3}{2}}\,\text{particles/cm}^3. \qquad (5.73)$$

All metals will have evaporated long before this limiting temperature T_L is reached. Only the semimetals, with their much lower carrier density, offer a possibility of lifting the degeneracy; for bismuth, for instance, with an m^*/m of $0\cdot1$ and a D of 5×10^{16}, the degeneracy is lifted at $60°\,K$.

Having verified that the conduction electrons of all metals are highly degenerate, we want to calculate the effect which is most directly indicative of their degenerate condition, namely their *heat capacity*. Let us introduce the Fermi function $F(x)$ as

$$F(x) = \frac{1}{e^x + 1}, \qquad (5.74)$$

then the total number N of conduction electrons equals

$$N = \int_{-\infty}^{+\infty} g(\epsilon)\, F\left(\frac{\epsilon - \eta}{kT}\right)\,d\epsilon. \qquad (5.75)$$

The total energy E of these electrons is

$$E = \int_{-\infty}^{+\infty} \epsilon g(\epsilon) F\left(\frac{\epsilon-\eta}{kT}\right) d\epsilon. \qquad (5.76)$$

The quantity to be computed is the heat capacity of the electron assembly, that is the temperature derivative of (5.76). In taking this derivative we must assume η to be a function of T, such that the integral (5.75) for N remains constant. This means that we have

$$C = \left(\frac{\partial E}{\partial T}\right)_N = \left(\frac{\partial E}{\partial T}\right)_\eta - \frac{1}{(\partial N/\partial \eta)_T}\left(\frac{\partial E}{\partial \eta}\right)_T\left(\frac{\partial N}{\partial T}\right)_\eta, \qquad (5.77)$$

which, from (5.75) and (5.76), becomes

$$C = -\frac{1}{kT^2}\int_{-\infty}^{+\infty} \epsilon(\epsilon-\eta)g(\epsilon)F_x'\left(\frac{\epsilon-\eta}{kT}\right)d\epsilon$$

$$+\frac{1}{kT^2}\frac{\displaystyle\int_{-\infty}^{+\infty}\epsilon g(\epsilon)F_x'\left(\frac{\epsilon-\eta}{kT}\right)d\epsilon \int_{-\infty}^{+\infty}(\epsilon-\eta)g(\epsilon)F_x'\left(\frac{\epsilon-\eta}{kT}\right)d\epsilon}{\displaystyle\int_{-\infty}^{+\infty}g(\epsilon)F_x'\left(\frac{\epsilon-\eta}{kT}\right)d\epsilon};$$

here F_x' is the derivative of the function (5.74). To simplify this observe that if the factors ϵ in the first term and in the first integral of the second term are replaced by $\epsilon-\eta$, equal contributions are added and subtracted. The substitution will be made; at the same time the argument $x = (\epsilon-\eta)/kT$ is conveniently introduced. This yields

$$C = -k^2T\int_{-\infty}^{+\infty}g(\eta+xkT)x^2F_x'\,dx + k^2T\frac{\left[\displaystyle\int_{-\infty}^{+\infty}g(\eta+xkT)xF_x'\,dx\right]^2}{\displaystyle\int_{-\infty}^{+\infty}g(\eta+xkT)F_x'\,dx}.$$

Because of the presence of F_x' in all integrals the contributions to C come exclusively from the neighbourhood of the Fermi level, as one would expect physically. Expansion of $g(\eta+xkT)$ in powers of x yields automatically an expansion of the specific heat in powers of kT/η. Since this is a small quantity we would expect to get a close approximation by computing the first term only, which takes $g(\epsilon)$

to be a constant. In this approximation the second term vanishes by symmetry and we get

$$C = g(\eta) k^2 T \left[-\int_{-\infty}^{+\infty} x^2 F'_x \, dx \right].$$

The integral in brackets is easily calculated. First we observe from (5.74) that the integrand is even. Hence the integral equals twice the integral from 0 to ∞. An integration by parts then yields for it

$$4 \int_0^\infty x F(x) \, dx = 4 \int_0^\infty x \, dx [e^{-x} - e^{-2x} + e^{-3x} - \ldots] = \frac{\pi^2}{3}.$$

The electronic heat capacity is therefore equal to

$$C = \frac{\pi^2}{3} g(\eta) k^2 T. \tag{5.78}$$

As usual, the quasi-free electron model with an effective mass is useful for an estimate. We may then substitute $g(\eta)$ from (5.68). For one mole of electrons we find

$$C \approx R \ \tfrac{1}{2}\pi^2 \frac{kT}{\eta}. \tag{5.79}$$

The expression differs from R in order of magnitude because of the small factor kT/η. It also has a temperature dependence which is easily distinguished from the T^3 type and from constant contributions arising from classical degrees of freedom. In specially favourable cases, such as the alkali metals, equation (5.79) can actually be used to determine the effective mass of the electrons; for there is little doubt in that case that there is one conduction electron per atom; so η results from (5.79) and m^* is then the only unknown in (5.71). Values of m^* differing very little from m are the result, particularly for sodium. In the case of more complicated metals with involved $W(\mathbf{k})$ curves the simplifications underlying (5.79) are not justified the measurement of C then yields only $g(\eta)$ through (5.78).

    With respect to outside bodies the degenerate electron gas inside a metal is characterized by the height of its Fermi level η which is also the chemical potential of the electrons. In equilibrium this quantity must of course be the same everywhere. Under normal operating conditions this equality of η between metals in electrical contact is reached through the deposition of surface contamination of various kinds and the flow of small static charges; it is therefore of little fundamental interest. However, it is possible to clean metal surfaces sufficiently well to avoid these accidental conditions. The η's of such clean surfaces are no longer equal, but the Fermi level of each metal lies by a characteristic amount below the level of an electron which is at rest at an atomically large distance outside the metal surface in vacuum. This amount W is called the *work function*; it is characteristic of the metal and the crystal face exposed. In many simple treatments, the quantity is treated as characteristic

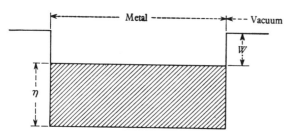

Fig. 5.12. Idealization of a metal as a 'lake' of degenerate free electrons. The level of the lake below the vacuum level is the work function W.

of the metal alone. The metal appears then as a 'lake' of degenerate electrons whose surface (Fermi level) is offset with respect to vacuum by a fixed amount W, as shown in fig. 5.12. The picture may be approximately valid in some cases, but in the well studied case of tungsten, the W so defined is not a constant of the material, but depends on the clean crystal face exposed. If several faces are simultaneously exposed, most of them will have electric double layers to adjust to some common value of the level difference W.

As long as we do not think of W as an absolute material constant, we may employ fig. 5.12 for qualitative reasoning. It indicates that the emergence of electrons from the metal surface is similar to evaporation, and that there will be a definite density of electrons outside the metal in equilibrium with the 'lake' inside. The density can be calculated from the equations (5.64) and (5.67). The most characteristic aspect of this density d is the amount by which the level $\epsilon = 0$ is larger than η, that is, our constant W; hence we may write

$$d = d_0 \, e^{-W/kT}.$$

A similar expression will result for the current I which can be drawn from the metal surface into vacuum:

$$I = I_0 e^{-W/kT}. \tag{5.80}$$

Equation (5.80) is called the *Richardson law*. The factor I_0 in it is not quite a constant, but it varies sufficiently slowly with temperature to allow the determination of W from semilog plots of I versus $1/T$ (Richardson plots). The values of W obtained in this way for metallic elements are generally similar to the ionization potential of the individual atoms constituting the metal.

TIGHT BINDING THEORY OF ELECTRONS IN SOLIDS

Up to this point the chapter has acquainted the student with Bloch functions and their characteristics. The moment has now come to warn him that these functions are not always the right ones or even the best ones. The Bloch functions are the logical outcome of the decoupling assumption made at the beginning of the chapter. This assumption is not nearly as well justified here as it is in the case of atoms; for the charge 'seen' by an electron wandering from atom to atom is comparable with its own charge, and the progress of one electron will almost always be coupled with movements compensating its charge in the vicinity. *Correlation* between electrons is therefore important. Let us therefore reassess the electronic wave functions of crystals, starting from the atoms, ions or molecules which compose it.

The simplest picture which one can form of crystals is to consider them an aggregate of almost undisturbed constituents. In some cases, such as the ionic crystals, this is not a bad viewpoint. In this approach, the stationary states of the electrons are taken over from atomic theory, and the influence of the crystalline condition upon these states is considered a perturbation. This point of view is the most generally useful. In fact, all other approaches accept this particular viewpoint for the inner core electrons, which form the vast majority of electrons in any crystal. The usefulness of this point of view can be enhanced by allowing small modifications which take into account the crystalline medium. For example, an atom or ion in a solid is embedded in a non-spherical environment. This removes the angular momentum quantum number J as a good quantum number, and splits levels of given J into sets of sublevels. These sublevels can be predicted in a straightforward way from the symmetry properties of an eventual non-spherical electric field. The interested reader will find some of these cases indicated in the problems of Chapter 3. It is more instructive for the purpose of this chapter to apply these basic ideas to the valence electrons, because the modifications imposed by the medium are for them of greater significance.

The tight binding theory of electrons considers the wave functions of electrons in crystals prescribed in principle by the theory of atomic structure. These states can thus be characterized as $1s$, $2s$, $2p$, $3s$, $3p$, $3d$, $4s$, etc. Let us take as illustrative examples hydrogen metal, a hypothetical substance which perhaps exists at high pressure and low temperature, and solid helium, a cubic close-packed structure which is known to exist under these conditions. According to the tight binding theory, we accept the wave functions of the electrons about each atom as the basic wave functions of the problem. We have thus a network of wave functions of the form

$$\psi_{n,r} = \psi_n(\mathbf{x} - \mathbf{r}), \qquad (5.81)$$

where n subsumes the atomic quantum numbers n, l, m, and $\mathbf{r}$ may take all values (1.2) corresponding to a face-centred cubic lattice. Of special interest among the states which can be formed from the wave functions (5.81) is the crystalline ground state and states in its immediate vicinity; they are presumably made up from ground state wave functions of the type (5.81) for each electron, since the first excited states of hydrogen and helium lie quite high; let us denote the ground state wave function by $n = 0$. A possible low lying many-electron wave function might then be given the form

$$\Psi(1, 2, ..., N) = \left\{ \prod_{i=1}^{N} \psi_0(\mathbf{x}_i - \mathbf{r}_i) \right\} \{\text{a spin function}\}, \qquad (5.82)$$

where the index i is such that, simultaneously, $\mathbf{x}_i$ runs over all electron co-ordinates and $\mathbf{r}_i$ over all lattice vectors; in hydrogen each lattice vector is taken once, in helium twice. The spin function need not be fully determined here. One can satisfy the anti-symmetry requirement for electrons by generalizing (5.82) and writing

$$\Phi(1, 2, 3, ..., N) = \sum_P (-1)^P P\{\Psi(1, 2, ..., N)\}. \qquad (5.83)$$

Here P represents an operator permuting the cartesian and spin co-ordinates of the electrons in Ψ, $\sum_P$ represents the sum over all such permutations, and $(-1)^P$ is a factor ± 1 according to whether the permutation is even or odd.

Although there is nothing obviously wrong with the wave function (5.83), there is an objection to it; we have used atomic wave functions to create something different in character without giving very much thought to this character. If we integrate the absolute square of the wave function Φ over all but one of the electron co-ordinates, say x_1, we get a probability density for that electron which is spread over the entire crystal and has the crystal period; this means that the electron must wander from atom to atom. If such a switching of atoms is a rare event then the wave function is presumably all right; for the electron is then circulating around one nucleus most of the time, and this feature is described correctly by the basis functions (5.81). However, if the switching of atoms is frequent, say comparable with the period of revolution around one atom, then new dynamical questions enter, which the basis wave functions do not take into account; this situation arises whenever the overlap of the atomic functions from atom to atom is large. One has then a situation in which there is effective one-electron motion through the crystal. This motion will be governed by an effective one-particle hamiltonian, yielding an effective one-particle Schrödinger equation. The solutions of such an equation fall under the sway of the general group theoretical theorem of p. 62. They cannot have the asymmetric form (5.81) but must have a wave vector $\mathbf{k}$ obeying (3.37). It is the beauty of the tight binding method that it can embrace this idea within its own framework. From the basic set (5.81) we can construct wave functions of the form

$$b(\mathbf{x}; \mathbf{k}) = \sum_\mathbf{r} i x p (\mathbf{k} \cdot \mathbf{r}) \psi(\mathbf{x} - \mathbf{r}), \qquad (5.84)$$

where $\mathbf{k}$ obeys (3.37). The wave functions (5.84) are the tight binding approach to the Bloch functions. We can obviously use them as well as the localized functions (5.81) to construct a many electron wave function. It would read, in analogy to (5.83)

$$\Phi = \sum_P (-1)^P P \left[\left\{ \prod_i b(\mathbf{x}_i; \mathbf{k}_i) \right\} (\text{a spin function}) \right], \quad (5.85)$$

where the index i is such that while $\mathbf{x}_i$ runs over all electron coordinates, $\mathbf{k}_i$ runs over an appropriate number of vectors of the first Brillouin zone.

Before going into the question whether (5.83) or (5.85) is a more appropriate wave function, we should realize first that in the case of solid helium they are identical. To show this, let us take a particular wave function (5.82) out of the sum (5.83), and within the spin function a particular spin product

$$\alpha(1)\,\alpha(2)\,\beta(3)\,\alpha(4)\,\beta(5)\dots\alpha(N).$$

Let us also assume that the electrons 1 and 2 have $\mathbf{r}_i = 0$. Now among the permuted terms in (5.83) there is one permuting the electrons 1 and 2. It cancels the term considered because both electrons have α spin. To get a non-vanishing result we should have picked a spin product of the form

$$\alpha(1)\,\beta(2)\,\beta(3)\dots,$$

whereupon the term would have combined with its permutation into

$$\psi(\mathbf{x}_1)\,\psi(\mathbf{x}_2)\,\{\alpha(1)\,\beta(2)-\beta(1)\,\alpha(2)\}\dots.$$

We may assume arbitrarily that (5.82) contains only

$$\psi(\mathbf{x}_1)\,\psi(\mathbf{x}_2)\,\alpha(1)\,\beta(2)$$

and then leave it to the permutations in (5.83) to deliver the second half of the expression. This argument may be continued to give to (5.82) the form

$$\begin{aligned}\Psi = \;&\psi(\mathbf{x}_1)\,\psi(\mathbf{x}_3-\mathbf{a})\,\psi(\mathbf{x}_5-\mathbf{b})\dots\\ &\times \psi(\mathbf{x}_2)\,\psi(\mathbf{x}_4-\mathbf{a})\,\psi(\mathbf{x}_6-\mathbf{b})\dots\\ &\times \alpha(1)\,\alpha(3)\,\alpha(5)\dots\beta(2)\,\beta(4)\,\beta(6)\dots. \quad (5.86)\end{aligned}$$

In combining permutations of (5.86) to form Φ it is useful to form first an intermediate which permutes only the even electrons among themselves, and the odd ones among themselves. This intermediate wave function Φ^+ reads

$$\Phi^+ = \begin{vmatrix} \psi(\mathbf{x}_1) & \psi(\mathbf{x}_3) & \psi(\mathbf{x}_5) & \cdots & \psi(\mathbf{x}_{2N-1}) \\ \psi(\mathbf{x}_1-\mathbf{a}) & \psi(\mathbf{x}_3-\mathbf{a}) & \psi(\mathbf{x}_5-\mathbf{a}) & \cdots & \psi(\mathbf{x}_{2N-1}-\mathbf{a}) \\ \psi(\mathbf{x}_1-\mathbf{b}) & \cdots & \cdots & \cdots & \cdots \\ \hdotsfor{5} \end{vmatrix} \times$$

$$\times \begin{vmatrix} \psi(\mathbf{x}_2) & \psi(\mathbf{x}_4) & \cdots & \cdots & \psi(\mathbf{x}_{2N}) \\ \psi(\mathbf{x}_2-\mathbf{a}) & \cdots & \cdots & \cdots & \cdots \\ \psi(\mathbf{x}_2-\mathbf{b}) & \cdots & \cdots & \cdots & \cdots \\ \hdotsfor{5} \end{vmatrix} \times$$

$$\times \alpha(1)\,\alpha(3)\ldots\alpha(2N-1)\,\beta(2)\,\beta(4)\ldots\beta(2N). \tag{5.87}$$

Finally we can permute electron co-ordinates between the even and odd set to get the complete wave function. The essence is that the complete wave function as well as (5.87) contains the one-electron wave functions ψ in two determinants of identical form. This type of determinant remains unaltered if we replace the wave functions by linear combinations. Hence introduction of the Bloch type wave functions (5.84) in the place of the atomic functions (5.81) leaves the final wave function Φ unaltered.

We see from this that the use of atomic wave functions reproduces automatically the phenomenon of chemical saturation. It is to a large extent meaningless to ask what the exact nature of the one-electron wave functions are within a chemically saturated crystal such as helium; for the final many-electron wave function partakes of apparently mutually exclusive descriptions. The question whether the wave functions (5.81) or (5.84) form the correct base is thus relevant only in our example of metallic hydrogen, the prototype (if existing) of a chemically unsaturated solid. In such a discussion it is not automatically clear that (5.84) is better. For it is seen that the wave functions derived from (5.81) assign only one electron to one proton, while in (5.85) there is a high probability of several electrons being near the same proton, because the distribution of each electron in the lattice is random. In short, the question has no clear-cut answer. If the interatomic distance is large, individual electrons tend to stay near their ion core and to exchange places in pairs; this is expressed in the hypothetical wave function (5.83). If the distances are small and the overlap of wave functions appreciable, electrons tend to move about independently under the influence of an effective hamiltonian having the crystal symmetry; this is expressed in the wave function (5.84) and other Bloch functions. Experience favours them for the outer electrons of the metals. For the inner shells they are definitely inappropriate. Between these two extremes, there arise intermediate cases such as the $3d$ shell of the transition metals and the $4f$ shell of the rare earths where neither of the two descriptions is fully adequate.

PROBLEMS

1. Prove the following three propositions for the δ-well model:
(a) All allowed bands are separated by forbidden gaps.
(b) As the separation d of δ-wells of fixed strength η is decreased the midpoint of the lowest band $\{k = (1/4d)\}$ rises.
(c) In the change discussed under (b), the energy of an isolated δ-well always remains an allowed energy.
2. Use (5.31) and (5.32) to give an estimate of the width of the lowest allowed band in the δ-well model when that width is small.
3. Prove the perturbation formula (5.42) for band edges in weak periodic fields directly for the δ-well model, utilizing (5.28).
4. Derive the formula analogous to (5.28) for the δ-barrier model. The model is another limiting case of the Kronig–Penney model, in which there is no potential except for very thin, infinitely high, potential barriers located at periodic intervals. The product of thickness and height is supposed to remain finite.
5. Suppose we have a two-dimensional periodic potential which in rectangular cartesian co-ordinates equals $\mathscr{V}(x, y) = V(x, d) + V(y, d)$ where $V(z, d)$ is a δ-well potential of strength $\eta = 1$ and variable period d. Carry out the following operations:
(a) Determine the minimum energy E_{min} and the maximum energy E_{max} in the lowest band as functions of d and the reduced wave vectors $\mathbf{k}_{min}$ and $\mathbf{k}_{max}$ which go with them.
(b) Prove that in the neighbourhood of E_{min} or E_{max} the energy varies linearly with $(\mathbf{k}-\mathbf{k}_{min})^2$ and $(\mathbf{k}-\mathbf{k}_{max})^2$ respectively (effective mass approximation); find the effective mass at either end.
(c) Find out whether proposition (b) also holds for the next higher band.
(d) Construct for some fixed d the contour in the k_x–k_y-plane for which the energy is the arithmetic mean of the energies calculated in (b). Deduce from it qualitatively the appearance of other energy contours in the lowest band.

6. Prove that unless two Bloch functions are degenerate in **k** as well as in energy, any pair of them obeying (5.44) are orthogonal to each other, even if the Born–von Karman conditions are not allowed.

7. Why are the operators (5.39) called translation operators?

8. Derive the equation (5.59) for the expectation value of the velocity for a Bloch function by using the quantum mechanical operator methods rather than wave packets. Prove therefore that the expectation value of the operator $-i\dfrac{\hbar}{m}\dfrac{\partial}{\partial x}$ for a Bloch function is $\dfrac{1}{h}\dfrac{\partial W}{\partial k_x}$.

9. Derive the equation (5.68) for the density of energy states of an electron gas from the statistical identity (5.67). Use it to derive the mean energy of an electron in an electron gas at absolute zero. Compare the mean energy with the Fermi energy η.

10. Derive the pressure of a Fermi gas at absolute zero
    (a) from the result of problem 9;
    (b) from kinetic or virial considerations applicable to all gases.

11. Can you supplement the result of problem 10 and derive the pressure of a Fermi gas at low temperature from the specific heat (5.79)?

BIBLIOGRAPHY

GENERAL BACKGROUND

F. SEITZ. *Modern Theory of Solids*. (New York, McGraw-Hill, 1940.)

R. E. PEIERLS. *Quantum Theory of Solids*. (Oxford, Clarendon Press, 1955.)

J. C. SLATER. The electronic structure of solids. *Handb. d. Phys.* **19**, 1–136. (Springer, Berlin, 1956.)

J. R. REITZ. Methods of the one-electron theory of solids. *Solid State Physics*, **1** (1955), 1–95.

MATHIEU FUNCTIONS

N. W. McLACHLAN. *Theory and Application of Mathieu Functions*. (Oxford, Clarendon Press, 1947.)

J. MEIXNER and F. W. SCHÄFKE. *Mathieusche Funktionen und Sphäroidfunktionen mit Anwendungen auf physikalishe und technische Probleme*. (Berlin, Springer, 1954.)

ELECTRONIC PROPERTIES OF METALS

N. F. MOTT and H. JONES. *The Theory of the Properties of Metals and Alloys*. (Oxford, Clarendon Press, 1936.)

A. SOMMERFELD and H. BETHE. Elektronentheorie der Metalle. *Handb. d. Phys.* **24**, part 2, pp. 333–622. (Berlin, Springer, 1933.)

A. H. WILSON. *The Theory of Metals*, 2nd ed. (Cambridge, University Press, 1953.)

ELECTRON EMISSION BY SOLIDS

C. HERRING and H. M. NICHOLS. Thermionic emission. *Rev. Mod. Phys.*
    21, (1949), 185–270.
W. B. NOTTINGHAM. Thermionic emission. *Handb. d Phys.* **21**, (1956).
    1–231.
R. H. GOOD, JR., and E. W. MÜLLER. Field emission. *Handb. d. Phys.*
    21 (1956), 176–231.

CHAPTER 6

ELECTRONS IN CHEMICALLY SATURATED SOLIDS; SEMICONDUCTORS

In Chapter 5 it was stated that the band theory of solids naturally classifies solids into conductors and insulators, according to whether their bands are all filled or not. It would have been more exact to say that the classification results for exactly stoechiometric solids at the absolute zero of temperature. If a compound is an insulator on this criterion, but is not at absolute zero, or if it has some impurity contained in it, a small concentration of electric carriers may be present in the specimen. A chemically pure material, which would be an insulator on the simple definition, may in fact be an *intrinsic semiconductor* at room temperature because a number of electrons from the highest filled band are thermally excited to the lowest empty band. This is not the only type of semiconductor possible. There exist so-called *donor* impurities, that is impurities which make one electron available for conduction; the result is an *excess* or *n-type semiconductor*. An example of this type of conductor is germanium containing a small amount of a pentavalent impurity such as phosphorus or arsenic. When these substances enter the four co-ordinated covalently bonded germanium lattice substitutionally they have one surplus electron. This electron is easily split off from the parent atom and wanders through the lattice. As long as the concentration of these electrons is kept low they will act more or less independently and thus follow more closely the original ideas of Drude than metallic electrons do. The reverse situation, the creation of a *defect* or *p-type semiconductor* through the presence of *acceptors* is also possible. These acceptors cause electrons to be missing in the highest filled band and thereby create *defect electrons* or *holes* in that band, which act as mobile positive carriers. The most easily understood acceptor centres are trivalent impurities such as boron atoms which enter substitutionally into a normal four co-ordinated position; here, they find themselves short of an electron for bonding. The defect can be supplied by one of the neighbours and thus can propagate itself through the lattice; in the course of this the impurity accepts an electron from the crystal and thus is termed an acceptor impurity. Vacancies in the crystal structure are often capable of holding an electron, and thus form another common type of acceptor centre. It is found empirically that the concepts of semiconductor electronics are extremely straightforward, at least in some materials. This straightforwardness has permitted the development of a theoretical technology, in which semiconductor devices are designed theoretically and their performance predicted. This chapter will treat this aspect of semiconductor physics only incidentally.† Our concern here is mainly with the justification of these concepts in the framework of solid state theory. It goes without saying that a solid body is such a complicated thing that a justification of a concept applicable to it always remains a partial one. A reader so cautioned will find a relatively logical development of semiconductor theory in the rest of this chapter.

There is a straightforward way to fit the notion of excess and defect electrons into the one-electron theory of solids discussed in the last chapter. Taking our case of solid helium as an example, we may add an electron to the system by multiplying the wave function (5.87) with the wave function for an extra electron in an excited state and summing over all permutations:

$$
\Phi = \sum_P (-)^P P \begin{vmatrix} b(\mathbf{x}_1; \mathbf{k}_1) & b(\mathbf{x}_2; \mathbf{k}_1) & \dots & b(\mathbf{x}_N; \mathbf{k}_1) \\ b(\mathbf{x}_1; \mathbf{k}_2) & \dots & \dots & b(\mathbf{x}_N; \mathbf{k}_2) \\ \dots\dots\dots\dots\dots\dots\dots\dots\dots\dots\dots\dots\dots \\ b(\mathbf{x}_1; \mathbf{k}_N) & \dots & \dots & b(\mathbf{x}_N; \mathbf{k}_N) \end{vmatrix} \times
$$

$$
\times \begin{vmatrix} b(\mathbf{x}_{N+1}; \mathbf{k}_1) & \dots & b(\mathbf{x}_{2N}; \mathbf{k}_1) \\ \dots\dots\dots\dots\dots\dots\dots\dots\dots \\ b(\mathbf{x}_{N+1}; \mathbf{k}_N) & \dots & b(\mathbf{x}_{2N}; \mathbf{k}_N) \end{vmatrix} \times B(\mathbf{x}_{2N+1}; \mathbf{k}_1)
$$

$$
\times \alpha(1)\alpha(2)\dots\alpha(N)\beta(N+1)\dots\beta(2N)\alpha(2N+1). \quad (6.1)
$$

Here the permutation operator P goes only over those permutations not absorbed into the determinants. B is an excited Bloch function constructed from excited atomic functions by the method (5.84).

---

† For fuller discussions of this aspect see the literature list at the end of the chapter.

The method can be made more flexible by dropping the insistence on using a single atomic function to construct B. Since these atomic functions are complete, we can in fact allow for B any function, except one of the b's (in the latter case Φ vanishes, because the exclusion principle is violated).

The same straightforward procedure can be used to describe a missing electron in the $1s$ shell of one of the helium atoms. We write

$$\Phi = \sum_P (-)^P P \begin{vmatrix} b(\mathbf{x}_2;\mathbf{k}_2) & b(\mathbf{x}_3;\mathbf{k}_2) & \dots & b(\mathbf{x}_N;\mathbf{k}_2) \\ b(\mathbf{x}_2;\mathbf{k}_3) & \dots & \dots & b(\mathbf{x}_N;\mathbf{k}_3) \\ \hdotsfor{4} \\ b(\mathbf{x}_2;\mathbf{k}_N) & \dots & \dots & b(\mathbf{x}_N;\mathbf{k}_N) \end{vmatrix}$$

$$\times \begin{vmatrix} b(\mathbf{x}_{N+1};\mathbf{k}_1) & \dots & b(\mathbf{x}_{2N};\mathbf{k}_1) \\ \hdotsfor{3} \\ b(\mathbf{x}_{N+1};\mathbf{k}_N) & \dots & b(\mathbf{x}_{2N};\mathbf{k}_N) \end{vmatrix}$$

$$\times \alpha(2)\,\alpha(3)\dots\alpha(N)\beta(N+1)\,\beta(N+2)\dots\beta(2N). \quad (6.2)$$

The wave function represents a defect electron of wave vector $\mathbf{k}_1$ and spin α. Clearly, if we deal with atoms having several closed shells, the defect can be placed in any of the shells. In practice, it is almost always in the filled band of highest energy, the so-called *valence band*. Similarly, the description (6.1) is used most often for an electron in the lowest empty band which is called the *conduction band*.

The equations (6.1) and (6.2) show that the invariance of Φ with respect to the nature of the basic one-electron wave functions no longer holds as soon as an electron is added to or removed from a complete shell. We therefore have to make a decision concerning the nature of the basic wave functions. Theory insists that this decision can only go one way: they must be Bloch functions. For the functions (6.1) and (6.2) supposedly satisfy a Schrödinger equation (at least approximately), and therefore are subject to the theorem of p. 62. This theorem is satisfied for the wave functions as written: a shift of all electron co-ordinates $\mathbf{x}_i$ by a lattice vector $\mathbf{r}$ satisfying (1.2) multiplies them with $\exp(-\mathbf{k}_1 \cdot \mathbf{r})$. If atomic functions are substituted in (6.2), and similarly for B in (6.1) the theorem is no longer obeyed.

Since we are forced to make a definite choice for the wave functions of the excess and defect electrons, the result of our choice must be appraised in the light of previous criticism. The first and most formidable criticism arises from scattering. There are materials whose bands are so narrow, and hence, from (5.59), whose electron speeds are so small, that the time to traverse an interatomic distance is comparable with the electronic mean free time between collisions. Now the band structure and the existence of a **k** vector arises from the lattice periodicity. If an electron is scattered in a time comparable with its time of flight for one lattice period then it cannot experience that period, and **k** is not a good quantum number. Experience indicates that this is the situation encountered in many oxides and sulphides; their conductivity is not zero, but is connected with an activation energy.† The proper picture seems to be that the normal transit of the electron from cell to cell by tunnelling is so slow that it cannot compete with the transit over thermally activated states. These activated states may be either activations of the lattice to configurations favourable for electron transit, or activation of the electron itself into a higher state which is less tightly bound. This would undoubtedly be the situation in our sample case of solid helium if a current could be measured at all. It follows from this analysis that the use of Bloch functions in these cases is irrelevant rather than wrong; the observed electron transport follows a more efficient path involving an activated state.

A second objection to the choice of a Bloch function for an excess or defect electron is the tight coupling of the extra particle with the electrons and ion cores of the saturated crystal. Authors who wish to stress this point may refer to such a particle complex as a 'polaron'. At one time it was thought even that an electron might be self-trapped in the medium which it polarizes. This surmise is incorrect.‡ The Bloch character of an excess or defect electron can be proved even under the condition of tight coupling. It becomes thereby independent of the one-electron theory of the last chapter. Let us

---

† R. R. Heikes and W. D. Johnston, *J. Chem. Phys.* **26** (1957), 582. J. Yamashita and T. Kurosawa, On the electronic current in NiO, *J. Phys. Chem. Solids*, **5** (1958), 34.

‡ C. Herring, *Phys. Rev.* **52** (1937), 356

re-derive this Bloch character for such electrons, basing ourselves on their low spatial density, rather than the one-electron approximation for atoms, molecules and crystals.

QUANTUM DESCRIPTION OF SEMICONDUCTOR ELECTRONS

The easiest place for a new start in describing semiconductor electrons is provided by excess semiconductors. The molar concentration of excess electrons in these materials is usually so small (10^{-5} to 10^{-9}) that they are effectively independent of each other because of their distance, even though they may be tightly coupled to the electrons of neighbouring atoms. In addition, for the same reason of distance, the presence of the positively charged donor atoms in the crystal can be ignored to a first approximation. The basic dynamical unit requiring study is therefore one excess electron, placed in a perfect, chemically saturated, insulating crystal. The assumption for the following treatment is that the presence of this electron does not destroy the stable non-degenerate ground state of the crystal in a fundamental way. In matrix language this means, if we use as a dynamical description the quantum states of the crystal with the electron held fixed at an arbitrary point $\mathbf{x}$, that for every $\mathbf{x}$ a ground state g is clearly discernible for the crystal, and that the matrix elements connecting these states for different values of $\mathbf{x}$ are large compared to the ones connecting these states to excited states of the crystal. This assumption is clearly satisfied if the one-electron approximation of the previous chapter is applicable. The inverse is not the case however. The electrons of the crystal may very well be tightly coupled to each other and to the extra electron, and nevertheless a description of the compound system by the co-ordinate of the extra particle may be possible.

If the assumption just formulated is correct, then we may get an approximately valid description of the motion of the electron by diagonalizing only the partial matrix block containing the elements referring to the crystal ground state g. The equation to be solved is then

$$\int \langle g, \mathbf{x} | \mathcal{H} | g, \mathbf{x}' \rangle \, d\mathbf{x}' \langle \mathbf{x}' | = E_g \langle \mathbf{x} |. \tag{6.3}$$

Because of the translational symmetry of the crystal the matrix elements above obey the relation

$$\langle g, \mathbf{x} + \mathbf{r} | \mathcal{H} | g, \mathbf{x}' + \mathbf{r} \rangle = \langle g, \mathbf{x} | \mathcal{H} | g, \mathbf{x}' \rangle, \tag{6.4}$$

where r is any vector obeying (1.2). Because of this, $\langle x+r|$ is a solution of (6.3) as well as $\langle x|$ with equal eigenvalue. Hence the theorem of p. 62 is applicable to these eigenvectors, and we can write them in the form of Bloch functions

$$\langle x| = b_q(x;k). \tag{6.5}$$

These Bloch functions are equivalent to (6.1) if the one-electron approximation is valid. In other cases we have a generalization defining for us the band concept in a tightly coupled system.

It is natural, for the description of an excess electron, to pass from the description in terms of g, x to one in terms of its Bloch states, that is g, q, k. The hamiltonian then takes the form

$$\langle g,q; k \left| \mathscr{H} \right| g,q'; k'\rangle = \delta_{q,q'}\delta_{k,k'}W_q(k), \tag{6.6}$$

which is equivalent to (5.47). In the same way the relation (5.59), namely

$$\langle g,q;k \left| \frac{dx}{dt} \right| g,q;k\rangle = \frac{1}{h}\frac{\partial W_q(k)}{\partial k}, \tag{6.7}$$

can be rederived in the new description.

In this way we see that the band description is assured for excess electrons in semiconductors. The extension to defect electrons is a little more delicate because it is obviously not possible to localize them as one does a real particle. However, some localization is possible. Since the crystal is periodic an electron equivalent to a given one can be found in each primitive cell; hence defect wave packets can be constructed which reduce the missing charge essentially to one single cell. We are thus led to investigate the ground state of a crystal after an electron has been removed from a definite atomic shell and a definite primitive cell of the lattice. We assume again, as previously, that the crystal has sufficient stability so that it makes sense to assume it in the ground state, and to diagonalize the hamiltonian only with respect to the state of the defect. In this respect we have to specify for the defect electron some quantum number, q^+, in addition to the cell in which it is located, because there are several different types of electrons in each cell which we may decide to remove. This quantum number q^+ enumerates the number of ways an electron can be extracted from the cell. In distinction from the previous case, q^+ can only take a finite number of values, namely the number Q of electrons per cell. The result of evaluating the hamiltonian for such states will thus be a matrix with elements of the type

$$\langle q^+; \mathbf{r} | \mathcal{H} | q^{+'}; \mathbf{r}' \rangle.$$

We have therefore the problem of solving the equation

$$\sum_{q^{+'}, \mathbf{r}'} \langle q^+; \mathbf{r} | \mathcal{H} | q^{+'}; \mathbf{r}' \rangle \langle q^{+'}; \mathbf{r}' | = E \langle q^+; \mathbf{r} |, \qquad (6.8)$$

with the side condition

$$\langle q^+; \mathbf{r}+\boldsymbol{\rho} | \mathcal{H} | q^{+'}; \mathbf{r}'+\boldsymbol{\rho} \rangle = \langle q^+; \mathbf{r} | \mathcal{H} | q^{+'}; \mathbf{r}' \rangle \qquad (6.9)$$

arising from crystalline symmetry, for any $\boldsymbol{\rho}$ obeying (1.2).

Solution of equation (6.8) is immediate if there is only one index q^+. The solution is, with the abbreviation (3.12),

$$\langle q^+; \mathbf{r} | = \frac{1}{\sqrt{N}} \mathrm{ixp}\,\mathbf{k} \cdot \mathbf{r}, \qquad (6.10)$$

and the energy equals

$$E = \sum_{\mathbf{r}'} \mathrm{ixp}\,\mathbf{k} \cdot (\mathbf{r}' - \mathbf{r}) \langle q^+; \mathbf{r} | \mathcal{H} | q^+; \mathbf{r}' \rangle. \qquad (6.11)$$

The result is simply our usual group theoretical simplification in a new guise. If there are several states q^+ from which the electron might be missing, then the transformation (6.10) can at least be used to make the problem finite; the procedure is familiar to the reader from the case of the phonons (equation (3.35)). The transformation resulting from (6.10) for an arbitrary operator is

$$\langle q^+; \mathbf{k} | \mathcal{S} | q^{+'}; \mathbf{k}' \rangle = \frac{1}{N} \sum_{\mathbf{r}, \mathbf{r}'} \mathrm{ixp}\,(\mathbf{k}' \cdot \mathbf{r}' - \mathbf{k} \cdot \mathbf{r}) \langle q; \mathbf{r} | \mathcal{S} | q'; \mathbf{r}' \rangle. \qquad (6.12)$$

Applying this to the energy we find

$$\langle q^+; \mathbf{k} | \mathcal{H} | q^{+'}; \mathbf{k}' \rangle = \left[\sum_{\mathbf{r}'-\mathbf{r}} \mathrm{ixp}\,\mathbf{k} \cdot (\mathbf{r}' - \mathbf{r}) \langle q^+; \mathbf{r} | \mathcal{H} | q^{+'}; \mathbf{r}' \rangle \right] \times$$
$$\times \left[\frac{1}{N} \sum_{\mathbf{r}'} \mathrm{ixp}\,(\mathbf{k}' - \mathbf{k}) \cdot \mathbf{r}' \right].$$

It is seen from (6.9) that the double sum is a product of two simple sums, one over $\mathbf{r} - \mathbf{r}'$, the other over $\mathbf{r}'$. The second sum was evaluated repeatedly before and yields $N\delta_{\mathbf{k},\mathbf{k}'}$. Hence

$$\langle q^+; \mathbf{k} \,|\, \mathscr{H} \,|\, q^{+'}; \mathbf{k}'\rangle = \delta_{\mathbf{k},\mathbf{k}'} \sum_{\mathbf{r}} i \mathrm{xp}\, \mathbf{k}\cdot\mathbf{r} \langle q^+; 0 \,|\, \mathscr{H} \,|\, q^{+'}; \mathbf{r}\rangle, \quad (6.13)$$

that is, the matrix is diagonal in $\mathbf{k}$. Since it was pointed out already that q^+ can only take a finite number of values (namely the number Q of electrons per primitive cell) we have for each $\mathbf{k}$ a finite matrix problem for which diagonalization is straightforward in principle. The result is Q eigenvalues $W(\mathbf{k})$. We can redefine q^+ (which was not completely specified so far) to label the Q different energy bands. We thus have

$$\langle q^+; \mathbf{k} \,|\, \mathscr{H} \,|\, q^{+'}; \mathbf{k}'\rangle = \delta_{\mathbf{k},\mathbf{k}'}\delta_{q^+,q^{+'}} W_{q^+}(\mathbf{k}). \quad (6.14)$$

This result, when compared with (6.6), expresses the far reaching similarity of excess and defect electrons in chemically saturated crystals. Derivation of equation (6.7) for a defect electron presents a formal complication, since only the cell number, not the co-ordinate can be specified for it; but the result is the same. In the application of the defect electron or hole concept, it is found that the distinction in the range between q and q^+ has no practical significance.

The discussion of the concept of a defect electron has furnished us incidentally with an interesting and useful transformation. The transformation is given by (6.12); it allowed us to transform an operator from a cell number base to a Bloch base. The transformation is the bridge between the Bloch functions and a localized description of a crystal. This local description is very convenient in many problems, not only because it is more intuitive, but also because many perturbations of crystals are given to us in local terms. In particular, the transformation must be applicable to the Bloch functions themselves; this yields a function which is as localized as possible for a wave function composed of Bloch functions from a single band. From (6.10) or (6.12) we get for this *cellular function* the formula[†]

$$a_q(\mathbf{x}; \mathbf{r}) = \frac{1}{\sqrt{N}} \sum_{\mathbf{k}} i \mathrm{xp}\,(-\mathbf{k}\cdot\mathbf{r})\, b_q(\mathbf{x}; \mathbf{k}).$$

[†] G. H. Wannier, *Phys. Rev.* **52** (1937), 191.

The chief property of $a_q(\mathbf{x};\mathbf{r})$ is that it depends on its two arguments only through their difference. Indeed, we get from (5.48) and the above, for a $\boldsymbol{\rho}$ satisfying (1.2),

$$a_q(\mathbf{x}+\boldsymbol{\rho};\mathbf{r}+\boldsymbol{\rho}) = \frac{1}{\sqrt{N}}\sum \mathrm{ixp}\,(-\mathbf{k}\cdot\mathbf{r}-\mathbf{k}\cdot\boldsymbol{\rho})\,b_q(\mathbf{x}+\boldsymbol{\rho};\mathbf{k})$$

$$= \frac{1}{\sqrt{N}}\sum_{\mathbf{k}} \mathrm{ixp}\,(-\mathbf{k}\cdot\mathbf{r})\,b_q(\mathbf{x};\mathbf{k})$$

$$= a_q(\mathbf{x};\mathbf{r}).$$

Hence we write the transformation equation in the form

$$a_q(\mathbf{x}-\mathbf{r}) = \frac{1}{\sqrt{N}}\sum_{\mathbf{k}} \mathrm{ixp}\,(-\mathbf{k}\cdot\mathbf{r})\,b_q(\mathbf{x};\mathbf{k}). \qquad (6.15)$$

The inverse of this unitary transformation reads

$$b_q(\mathbf{x};\mathbf{k}) = \frac{1}{\sqrt{N}}\sum_{\mathbf{r}} \mathrm{ixp}\,(\mathbf{k}\cdot\mathbf{r})\,a_q(\mathbf{x}-\mathbf{r}). \qquad (6.16)$$

A resemblance to the tight binding formula (5.84) is apparent in this equation. The advantage of the present treatment over the earlier one is that it is not restricted to a limiting case, and that the cellular functions a_q, as well as the Bloch functions b_q, form an orthonormal complete set. One can, in fact, exploit this similarity to make atomic functions $\psi_n(\mathbf{x}-\mathbf{r})$ orthonormal.† In analogy to the transformation pair (6.15) and (6.16), the operator transformation (6.12) can of course also be inverted, yielding

$$\langle q;\mathbf{r}|\mathscr{S}|q';\mathbf{r}'\rangle = \frac{1}{N}\sum_{\mathbf{k},\mathbf{k}'} \mathrm{ixp}\,(\mathbf{k}\cdot\mathbf{r}-\mathbf{k}'\cdot\mathbf{r}')\langle q;\mathbf{k}|\mathscr{S}|q';\mathbf{k}'\rangle. \quad (6.17)$$

A further step that can be usefully taken to widen the bridge between a localized picture and Bloch functions is to introduce the lattice vector $\mathbf{r}$ as a quantum operator. The obvious definition is

$$(\mathbf{r})_{\mathrm{op}}\,a_q(\mathbf{x}-\mathbf{r}) = \mathbf{r}\,a_q(\mathbf{x}-\mathbf{r}). \qquad (6.18)$$

Substitution of this into (6.16) yields

$$(\mathbf{r})_{\mathrm{op}}\,b_q(\mathbf{x};\mathbf{k}) = \frac{1}{2\pi i}\frac{\partial b_q(\mathbf{x};\mathbf{k})}{\partial \mathbf{k}}.$$

The proper operator equation for $\mathbf{r}$ results from this by treating

† This procedure is carried out in the preceding reference.

an arbitrary wave function Ψ as a superposition of such Bloch functions. In other words we write

$$\Psi(\mathbf{x}) = \sum_{q,\mathbf{k}} \alpha(q;\mathbf{k}) b_q(\mathbf{x};\mathbf{k}),$$

where $\alpha(q;\mathbf{k})$ is the amplitude function which characterizes the function Ψ. The result of the operator $\mathbf{r}$ on Ψ then takes the form

$$(\mathbf{r})_{\mathrm{op}}\Psi = \sum_{q,\mathbf{k}} \frac{\partial}{\partial \mathbf{k}}\left[\frac{1}{2\pi i}\alpha(q;\mathbf{k}) b_q(\mathbf{x};\mathbf{k})\right]$$
$$- \sum_{q,\mathbf{k}}\left[\frac{1}{2\pi i}\frac{\partial\alpha(q;\mathbf{k})}{\partial \mathbf{k}}\right] b_q(\mathbf{x},\mathbf{k}).$$

The first term on the right is zero, or rather becomes zero when the Born–von Karman volume is made infinite. The band becomes then a true continuum and the sum over $\mathbf{k}$ an integral; this integral is taken over a period for the derivative of a periodic function and thus vanishes.† The second term is the one desired, because it gives the result of the operator $\mathbf{r}$ on the amplitude function α. We find from it that

$$\langle q;\mathbf{k}|\mathbf{r}|q';\mathbf{k}'\rangle = \delta_{q,q'}\delta(\mathbf{k}-\mathbf{k}')\frac{i}{2\pi}\frac{\partial}{\partial \mathbf{k}}. \qquad (6.19)$$

This relation shows that $\mathbf{r}$ and $h\mathbf{k}$ are conjugate variables, and that we may write the symbolic relation

$$r_l k_j - k_j r_l \approx \frac{i}{2\pi}\delta_{l,j}, \qquad (6.20)$$

and in a representation in which $\mathbf{r}$ is diagonal

$$\mathbf{k} \approx \frac{1}{2\pi i}\frac{\partial}{\partial \mathbf{r}}. \qquad (6.21)$$

Some care must be used in the application of (6.20) and (6.21). It is known that from the relation (6.20) alone, quantum mechanics proves that the components of $\mathbf{r}$ and the components of $\mathbf{k}$ have a continuous spectrum from $-\infty$ to $+\infty$.‡ This is not true here,

† This passage to the limit is fully justified physically because the Born–von Karman volume is only a convenient crutch to write down certain formulas. A formulation assuming the crystal infinite throughout is possible, but slightly more awkward to write down.

‡ See, for instance, P. A. M. Dirac, *The Principles of Quantum Mechanics*, first edition (Oxford, Clarendon Press, 1930), p. 54.

since by the construction (6.18) $\mathbf{r}$ has the discrete spectrum (1.2), realized through the functions $a_q(\mathbf{x}-\mathbf{r})$. The error committed was to forget the indeterminacy of $\mathbf{k}$ in the reciprocal lattice. A properly defined function of $\mathbf{k}$ must be periodic in the reciprocal lattice, which means it must be a sum of crystal translation operators of the type (5.39). Both $b_q(\mathbf{x}; \mathbf{k})$ and $W_q(\mathbf{k})$ satisfy this condition by (5.49) and (5.50). $\mathbf{k}$ itself does not satisfy this condition and hence the direct use of $\mathbf{k}$ in (6.20) and (6.21) is not quite correct. A correct consequence of (6.20) is the following relation for the crystal translation operators

$$e^{2\pi i \mathbf{k} \cdot \rho}\,\mathbf{r} - (\mathbf{r}+\rho)\,e^{2\pi i \mathbf{k} \cdot \rho} = 0, \qquad (6.22)$$

where ρ is any vector satisfying (1.2). The new relation safeguards the indeterminacy of $\mathbf{k}$, and is thus an acceptable substitute for (6.20). We see from the relation (6.22) that by watching this indeterminacy aspect we have automatically safeguarded also the discrete spectrum of $\mathbf{r}$. Similar caution must be observed in the use of (6.21); differentiation with respect to a discrete variable is a risky procedure. However, if we restrict $\mathbf{k}$ to appear only through the intermediary of one of the crystal translation operators defined in (6.22) then we get generally

$$e^{2\pi i \mathbf{k} \cdot \rho} f(\mathbf{r}) = e^{\rho \cdot \frac{\partial}{\partial \mathbf{r}}} f(\mathbf{r})$$

$$= \left\{ 1 + \rho \cdot \frac{\partial}{\partial \mathbf{r}} + \frac{1}{2}\left(\rho \cdot \frac{\partial}{\partial \mathbf{r}}\right)^2 + \ldots \right\} f(\mathbf{r}),$$

or $\qquad\qquad e^{2\pi i \mathbf{k} \cdot \rho} f(\mathbf{r}) = f(\mathbf{r}+\rho)\,e^{2\pi i \mathbf{k} \cdot \rho}, \qquad (6.23)$

in accord with the discrete spectrum of $\mathbf{r}$. Thus, for allowable functions of $\mathbf{k}$, the statement (6.21) is correct.

As an example of the convenience of the operator method let us rederive the expectation value (5.59) or (6.7) of the velocity by finding the eigenvalues of $d\mathbf{r}/dt$:

$$\frac{d\mathbf{r}}{dt} = \frac{i}{\hbar}(\mathcal{H}\mathbf{r} - \mathbf{r}\mathcal{H}).$$

If we adopt a $q, \mathbf{k}$ representation, both $\mathcal{H}$ and $\mathbf{r}$ are seen to be diagonal in q, from the equations (6.6) and (6.18). (6.6) gives also the form of $\mathcal{H}$ to be used at fixed q; (6.19) gives the form of $\mathbf{r}$. We find that $d\mathbf{r}/dt$ is diagonal in q and $\mathbf{k}$ and equal to

$$\frac{d\mathbf{r}}{dt} = \frac{1}{h}\frac{\partial W_q(\mathbf{k})}{\partial \mathbf{k}}, \qquad (6.24)$$

in agreement with the result (6.7).

The reader will find, in the applications which are to follow, that the relation (6.21) will be used beyond the narrow interpretation (6.23). We are dealing then with an approximation. However, the approximation can be very good if $f(\mathbf{r})$ in (6.23) is of the right structure. The intention is to write something like

$$e^{\boldsymbol{\rho}\cdot\frac{\partial}{\partial\mathbf{r}}}f(\mathbf{r}) \sim \left[1+\boldsymbol{\rho}\cdot\frac{\partial}{\partial\mathbf{r}}+\frac{1}{2}\left(\boldsymbol{\rho}\cdot\frac{\partial}{\partial\mathbf{r}}\right)^2\right]f(\mathbf{r}). \qquad (6.25a)$$

This means that we try to employ a small number of differential operators to approximate the crystal translation operator on the left. Substitution of results into (6.25a) yields

$$f(\mathbf{r}+\boldsymbol{\rho}) \sim f(\mathbf{r})+\boldsymbol{\rho}\cdot\frac{\partial f}{\partial\mathbf{r}}+\frac{1}{2}\left(\boldsymbol{\rho}\cdot\frac{\partial}{\partial\mathbf{r}}\right)^2 f(\mathbf{r}), \qquad (6.25b)$$

which means approximation of a function by the first few terms of its Taylor series. This will be good if $f(\mathbf{r})$ is a slowly varying function.

In summary, (6.21) can be employed to approximate the crystal translation operators by differential operators if

(a) there is a natural starting point $\mathbf{k}_0$ about which the operator function $F(\mathbf{k})$ is developable in power series;

(b) all preliminary steps have been taken to make the function $f(\mathbf{r})$ on which $F(\mathbf{k})$ operates a slowly varying function of $\mathbf{r}$.

These two conditions will guide us in the application of (6.21). Its most obvious application is the transformation of the hamiltonian (6.6) to a cellular basis by writing

$$\langle q;\mathbf{r}|\mathcal{H}|q';\mathbf{r}'\rangle = \delta_{q,q'}\delta(\mathbf{r}-\mathbf{r}')W_q\left(-\frac{i}{2\pi}\frac{\partial}{\partial\mathbf{r}}\right). \qquad (6.26)$$

We shall see below that the necessary generalizations of (6.26) can be made to take care of common complexities in the band structure.

QUANTUM MECHANICS OF ELECTRONS IN SLOW FIELDS;
BOUND STATES

Up to this point the theory of excess and defect electrons in semi-conductors is the equivalent, for a crystalline medium, of a theory of particles in empty space. In other words, the periodic structure in which they are moving modifies their motion in some respects, but it leaves them in many ways 'free'; in particular in almost all quantum states the particles move at constant speed under the influence of no force. The theory is, however, of the framework type, that is, it corresponds up to now to an undisturbed crystal. The theory only becomes useful if it permits us to predict the reaction of the carriers to disturbances of the state described.

The reader is no doubt familiar with the way a physicist perturbs a physical system by *radiation* to elucidate its structure. The subject matter of Chapter 2 is a case in point; in that case attention was focused on geometrical structure. The use of radiation as a tool for elucidating the energy spectrum of electrons is illustrated in fig. 5.10, which shows part of the soft X-ray spectrum of aluminium. An example from a different part of the spectrum is paramagnetic resonance. In this technique a magnetized sample is irradiated by radiation whose quantum equals the level difference in equation (4.4); this radiation lies in the radiofrequency or microwave range. Just as in other fields of physics, a thorough discussion of a solid perturbed by radiation is not indispensable for an understanding of solid state physics, because radiation is primarily used to make a level structure visible which exists without it.

A more specific and important perturbation of electrons in solids are the *slowly varying electric fields* within it. They are associated with disturbances of the crystalline order. A well-understood case of this kind is the so-called *n-p*-junction in semiconductors. Such a junction can be produced artificially if a semiconductor single crystal is grown in such a way that the first half of it has a surplus of donors over acceptors, the second half a surplus of acceptors over donors. In such a crystal the position on an energy scale of the band system with respect to the Fermi level is not a constant. The distortion takes place in the band system and not the Fermi energy because the Fermi energy is the chemical potential of the electrons, which must be a constant throughout in equilibrium. Another way to see this is to start with the assumption that the band system is flat to begin with and the Fermi level distorted, in such a way that the electron density on the p-side is low. Then the excess electrons, seeing no opposition to their motion from the n- to the p-side, will start wandering over; at the same time defect electrons will wander the other way. This process sets up a charge unbalance and an internal field. To a first approximation this unbalance is an electric double layer between the ionized donors and acceptors and their

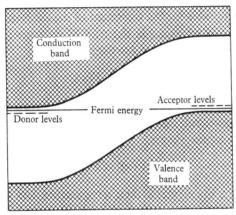

Fig. 6.1. Distortion of band energies in an n-p junction; the Fermi energy is the same throughout, and the bands are pinned on the two sides by local equilibrium conditions.

shifted carriers; this produces a shift in potential of one side with respect to the other. As soon as the unbalance has evened out the Fermi level, so that the same electron distribution (5.62) with the same η holds throughout, the motion will stop. Fig. 6.1 shows the resultant position of the band system on an energy scale. The abscissa is supposed to be a position co-ordinate running at right angles to the junction. On the left the semiconductor is n-type and has a surplus of donors over acceptors; on the right it has a surplus of acceptors and is p-type. The horizontal line going through the graph is the Fermi level for electrons. Its position at the extreme left is determined by the local form of the condition (5.63); in this case a balance must be struck between electrons on donor centres (shown by dashes) and electrons in the conduction band. A similar adjustment takes place on the right between electrons missing in the valence band and occupied acceptor centres (also shown by dashes). Within the junction, the electrons distribute themselves in a selfconsistent manner, so that they can obey the Fermi–Dirac law in the potential which they themselves produce. The result is the frozen-in electric field shown. Similar fields exist in the neighbourhood of impurities, voids, flaws, surfaces, etc. In some cases these frozen-in fields can be biased by applied fields to produce currents. A study of electron dynamics in such fields is thus of great importance. In addition, the theoretical possibilities are rather favourable for the treatment of the subject.

Let us take as an example the case of an excess electron; its hamiltonian can be written in the form

$$\mathcal{H} = \mathcal{H}_0 + V(\mathbf{x}). \qquad (6.27)$$

Here $\mathcal{H}_0$ is the undisturbed hamiltonian (6.6) previously denoted by $\mathcal{H}$. The potential $V(\mathbf{x})$ is assumed to be slowly varying with respect to the lattice, that is,

$$a\,|\boldsymbol{\nabla} V(\mathbf{x})| \ll V(\mathbf{x}). \qquad (6.28)$$

At every point, the crystal is then recognizable, but a local energy shift is imposed by the presence of $V(\mathbf{x})$. This local shift is ideally adapted to the cellular representation, because the variation of $V(\mathbf{x})$ within a cell has importance only in exceptional cases. We want therefore to consider matrix elements of the form

$$\langle q;\mathbf{r}\,|V|\,q';\mathbf{r}'\rangle = \int a_q^*(\mathbf{x}-\mathbf{r})\,V(\mathbf{x})\,a_{q'}'(\mathbf{x}-\mathbf{r}')\,d\mathbf{x}. \qquad (6.29)$$

Observing these matrix elements, we notice that because of (6.28) the off-diagonal ones are small; indeed develop V in a power series around some point $\mathbf{x}_0$:

$$V(\mathbf{x}) = V(\mathbf{x}_0) + (\mathbf{x}-\mathbf{x}_0)\cdot\boldsymbol{\nabla} V(\mathbf{x}_0) + \dots.$$

The first term has no off-diagonal elements at all, the second term gives an element of the type

$$\boldsymbol{\nabla} V(\mathbf{x}_0)\cdot\int a_q^*(\mathbf{x})\,a_{q'}(\mathbf{x}-\mathbf{r})\,(\mathbf{x}-\mathbf{x}_0)\,d\mathbf{x}.$$

The largest of these integrals arise from wave functions $a(\mathbf{x})$ located on the same cell or two neighbouring cells; $\mathbf{x}_0$ may then be chosen somewhere in the middle, and hence $\mathbf{x} - \mathbf{x}_0$ is of the order of a lattice distance; consequently the whole integral is at most of the order of a lattice distance, but probably much less. Thus, the off-diagonal elements are at most of the order of the left-hand side of (6.28), while the diagonal ones are of the order of the right. The former are thus negligible to a first approximation and the hamiltonian (6.27) can be given the approximate form

$$\mathcal{H} = W_q(\mathbf{k}) + V(\mathbf{r}). \tag{6.30}$$

Here we also omit the index q on $V_q(\mathbf{r})$ on the reasoning that the average value of $V(\mathbf{x})$ in one cell must be practically independent of the wave function employed in the averaging process.

The neatest applications of (6.30) arise in connection with impurity centres. In these cases, the potential $V(\mathbf{r})$ is approximately spherically symmetrical. If it is repulsive it can only be observed through scattering. In the attractive case there is the possibility of *bound states*; these states are of particular experimental interest if they lie in the forbidden energy gap. If the carriers return to their donors and acceptors they are, in fact, in such stationary states. In addition experiment has uncovered the existence of *traps*, that is attractive centres which, in the case of excess electrons, lie higher than the donor centres. According to the Fermi distribution law, very few of these centres hold electrons permanently; they can hold them for long times, however, and create thereby characteristic delay effects in the electric current. The most reasonable picture we can make of insulators today is, in fact, to assume them semiconductors which are so full of traps and junction barriers as to prevent the flow of excess carriers, even if they are injected from the outside.

The simplest model for such a centre is the one proposed by Koster and Slater.† This model assumes that $V(\mathbf{r})$ is zero everywhere except at $\mathbf{r} = \mathbf{0}$, at which point it has the value V_0. One asks for the energy eigenvalues of (6.30) for such a $V(\mathbf{r})$. The calculation goes easiest in the $\mathbf{k}$-representation in which $W_q(\mathbf{k})$ is diagonal. Since $V(\mathbf{r})$ is not an analytic function of $\mathbf{r}$ the differential methods

† G. F. Koster and J. C. Slater, *Phys. Rev.* **96** (1954), 1208.

evolved are not useful; instead, one gets the matrix elements for a particular band $q = 0$ directly from (6.12):

$$\langle \mathbf{k} \,|\, V \,|\, \mathbf{k'} \rangle = \frac{V_0}{N}. \tag{6.31}$$

The matrix equation resulting from (6.30) and (6.31) reads

$$\{W(\mathbf{k}) - E\} \, \phi(\mathbf{k}) + \frac{V_0}{N} \sum_{\mathbf{k'}} \phi(\mathbf{k'}) = 0. \tag{6.32}$$

where $\phi(\mathbf{k})$ is the amplitude with which the undisturbed solution $b_0(\mathbf{x}; \mathbf{k})$ contributes to the total solution. Equation (6.32) is solved by division by $W(\mathbf{k}) - E$, and summation over $\mathbf{k}$. The amplitudes $\phi(\mathbf{k})$ thereby drop out and we get an equation for the energy eigenvalues

$$f(E) = \sum_{\mathbf{k}} \frac{1}{W(\mathbf{k}) - E} + \frac{N}{V_0} = 0. \tag{6.33}$$

Equation (6.33) has N roots, because multiplication of $f(E)$ with $\prod_{\mathbf{k}} \{W(\mathbf{k}) - E\}$ yields a polynomial in E of degree N. All these roots are real. They are certainly real in the limit $V_0 = 0$ when they are the unperturbed values of $W(\mathbf{k})$ allowed by the Born–von Karman conditions. If V_0 is not zero we can locate the roots in certain intervals by consideration of $f(E)$. The function consists of a series of poles located at the unperturbed energy values. In between these poles $f(E)$ is continuous, being negative when E is slightly larger than a pole and positive when E is slightly smaller. In addition, $f(E)$ has a positive derivative in each of these ranges, as is seen by direct differentiation; it follows that there is one and only one root between any two roots of the unperturbed equation. In the limit of large E we have $f(\infty) = N/V_0$, that is, it is finite and has the sign of V_0. For positive V_0 there is therefore another sign change between the highest unperturbed energy state in the band and $+\infty$, and for negative V_0 between $-\infty$ and the lowest energy state of the band. A qualitative picture of $f(E)$ is shown in fig. 6.2. It is seen that the structure of the function locates all roots on the real axis, and all but one within the band. The roots are all smaller than the unperturbed roots if V_0 is negative, and larger if V_0 is positive.

An even more precise result follows from differentiation of (6.33) with respect to V_0:

$$\frac{dE}{dV_0} \sum_{\mathbf{k}} \frac{1}{\{W(\mathbf{k}) - E\}^2} - \frac{N}{V_0^2} = 0. \tag{6.34}$$

The derivative of E with respect to V_0 is intrinsically positive, which means that the shift of E is in the same sense as V_0 for all roots and any value of V_0. However, the magnitude of the shift remains small for all roots except one, because they have to remain between two

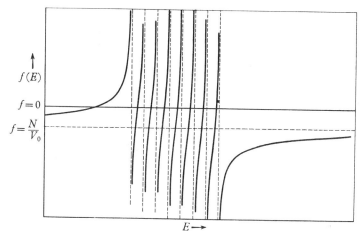

Fig. 6.2. Plot of $f(E)$ versus energy for the Koster–Slater model; the poles of $f(E)$ are the unperturbed levels; the roots are the perturbed levels.

successive eigenvalues of the quasicontinuous spectrum of the undisturbed band. Only one root can move by a really large amount. A rough estimate of this shift can be gained from the relation deducible from (6.33)

$$\sum E = \sum_{\mathbf{k}} W(\mathbf{k}) + V_0. \tag{6.35}$$

All roots together shift by the same amount as V_0, and since the sense of each shift is the same, the shift of E_0 is less than that of V_0. From (6.34), it is seen to be greater than 0. A tighter lower limit can be given by observing the maximum of the sum of the shifts of the levels within the band. Each of these levels is wedged between two successive unperturbed eigenvalues; it can at most move from one eigenvalue to the next one. The sum of these small shifts thus compounds at most to the sum of these intervals, that is the total band width W_0. The remainder of the shift (6.35) is thus attributable to the split off level E_0. We have thus the inequality

$$\frac{E_0}{V_0} > 1 - \frac{W_0}{|V_0|}, \tag{6.36a}$$

which together with the other two

$$E_0/V_0 < 1, \tag{6.36b}$$

$$E_0/V_0 > 0, \tag{6.36c}$$

gives a fair idea of the behaviour of E_0 as function of V_0 for an unspecified band shape. Among these inequalities the first two force the split-off level E_0 to shift by an amount about equal to the shift of V_0, if the magnitude of the latter is large compared to the band width W_0.

For more details the model has to be specialized further by specification of $W(\mathbf{k})$. We shall take here a one-dimensional example to continue. Suppose W is simply sinusoidal in k, occupying the energy values between 0 and W_0:

$$W(k) = \tfrac{1}{2} W_0 (1 - \cos 2\pi k d). \tag{6.37}$$

(6.33) then takes the form

$$\sum_{n=0}^{N-1} \frac{1}{\tfrac{1}{2} W_0 - E - \tfrac{1}{2} W_0 \cos (2\pi n/N)} + \frac{N}{V_0} = 0.$$

For the split-off level all denominators have the same sign and none of them tend to zero for large N. We can therefore pass to the limit $N = \infty$ which is the physically reasonable case. The equation becomes then

$$\frac{1}{2\pi} \int_{-\pi}^{\pi} \frac{d\phi}{\tfrac{1}{2} W_0 - E_0 - \tfrac{1}{2} W_0 \cos \phi} + \frac{1}{V_0} = 0, \tag{6.38a}$$

which solves to $E_0 = \tfrac{1}{2} W_0 + V_0 \{1 + (W_0^2/4 V_0^2)\}^{\frac{1}{2}}. \tag{6.38b}$

Equation (6.38) shows that E_0 as function of V_0 follows two half branches of a hyperbola. The common asymptote is $E_0 = V_0 + \tfrac{1}{2} W_0$, the apices are the band edges. A drawing of the result is given in fig. 6.3. The negative part of the diagram is particularly important.

In addition to furnishing an energy spectrum, the Koster–Slater model also gives an idea of the nature of the wave functions. We read off from (6.32) that

$$\phi(k) \propto \frac{1}{W(k) - E}.$$

This implies a strong preponderance of unperturbed states which are close in energy. For the levels inside the band, the preponderance is so great that in the limit $N = \infty$ the singularity is not even

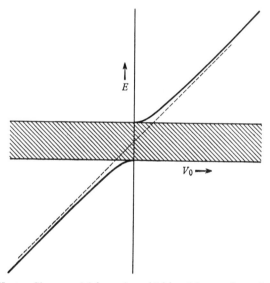

Fig. 6.3. Koster–Slater model for a sinusoidal band in one dimension. Energy plot for the band and the split-off level versus perturbing energy.

integrable, and the distribution becomes a δ-function in k, with the peak at the unperturbed k. For the split-off level the preponderance is not quite so great, but still present as long as V_0 is not comparable to the band width.

The Koster–Slater model is useful because it gives an example of the nature of the solutions of equation (6.30) without the introduction of further approximations. In general, we must think of $V(\mathbf{r})$ as a slowly varying function and insert (6.21) into (6.30) to get a manageable equation. The representation chosen is then a cellular representation and the amplitude function measures the amount of admixture of the various wave functions (6.15). Let us call this amplitude function $\phi_q(\mathbf{r})$. The equation for it reads then

$$W_q\left(-\frac{i}{2\pi}\frac{\partial}{\partial \mathbf{r}}\right)\phi_q(\mathbf{r}) + V(\mathbf{r})\,\phi_q(\mathbf{r}) = E\,\phi_q(\mathbf{r}). \qquad (6.39)$$

The equation begins to resemble a wave equation with a kinetic and a potential energy term. The analogy becomes perfect if the crystal is cubic and we deal with an energy minimum at $\mathbf{k} = 0$. We may infer for this case that the split-off bound states will be compounded largely from undisturbed states with small $\mathbf{k}$; this was pointed out in detail for the Koster–Slater model. The amplitude function $\phi_q(\mathbf{r})$ will then be generally slowly varying. The two conditions mentioned on p. 177 to make (6.25) applicable are then satisfied. Equation (6.39) is thus approximated by a second order differential equation whose exact form is determined by the structural relation (5.61). We find

$$-\frac{\hbar^2}{2m_q^*}\nabla^2\phi_q(\mathbf{r}) + V(\mathbf{r})\,\phi_q(\mathbf{r}) = \{E - W(0)\}\,\phi_q(\mathbf{r}). \qquad (6.40)$$

This is a Schrödinger equation in which the effective mass m_q^* replaces the true mass. This new type of mass thus does not only describe the relationship of velocity and momentum as discussed in Chapter 5, but it also enters in the kinetic energy term in an effective Schrödinger equation for the motion of such a particle in the presence of a disturbing field.

As an example of the usefulness of the equation (6.40) we shall consider the common case in which $V(\mathbf{r})$ originates from an impurity which acts as a point charge whose sign is opposite to that of the carrier. The donor and acceptor impurities mentioned earlier, which produce the carriers in the first place, are of this type. The potential resembles then an attractive Coulomb potential, and the medium enters into the picture only through the dielectric constant which reduces the attractive force by a factor $1/\epsilon$:

$$V(\mathbf{r}) = -\frac{e^2}{\epsilon\,|\mathbf{r}|}. \qquad (6.41)$$

Substitution of this expression into equation (6.40) yields the

Schrödinger equation for the hydrogen atom, and hence the energy levels

$$E_n = W_0 - \frac{e^4 m_q^*}{2\hbar^2 \epsilon^2 n^2} \qquad (6.42a)$$

with $\qquad n = 1, 2, 3, \ldots.$ $\qquad\qquad\qquad$ (6.42b)

This is a Balmer type of series with a reduced spacing because of the presence of the dielectric constant ϵ in the denominator. Critical perusal of the derivation shows that (6.42) is not likely to apply very often because a number of simplifying circumstances have to occur together to make it true. The only known case which is in agreement with this simple form of the theory is cuprous oxide.[†]

In most cases the equation determining the bound states of attractive centres in semiconductors does not quite simplify to yield a Balmer series. Little modification of the potential (6.41) has been introduced so far by the research workers; the value at $\mathbf{r} = 0$ is, of course, not right; Kohn and Luttinger found the distortion of the wave function at the seat of an acceptor atom so severe that they switched to a matching procedure for wave functions.[‡] Also, the question is debatable whether ϵ, which arises from the screening of intervening charges, is fully effective at short distances. Kohn[§] and Haken[||] did implement the standpoint adopted in this chapter by proving that ϵ enters as shown in (6.41) at sufficiently large distances.

More interesting than the modifications of the potential term is the handling of the first term of (6.30) for band structures which are not the simplest ones possible. The conduction bands of silicon and germanium[¶] do not have their energy minima at $\mathbf{k} = 0$. Consequently, the states splitting off the bottom of the band will vary approximately as the Bloch function going with the state at

[†] E. F. Gross, *Suppl. Nuovo Cimento*, **3** (1956), 672. R. J. Elliott, *Phys. Rev.* **108** (1957), 1384. Recent work (E. F. Gross, B. P. Zakharchenya and P. P. Pavinskii, *Journ. Tech. Phys.* **27** (1957), 2177 seems to indicate that the entity in cuprous oxide is an exciton, not a carrier in an impurity state. An exciton is an excess-defect electron pair which also satisfies the equations derived here; in fact, they were first derived for that unit (G. H. Wannier, *Phys. Rev.* **52** (1937), 191).
[‡] W. Kohn and J. M. Luttinger, *Phys. Rev.* **97** (1955), 883.
[§] W. Kohn, *Phys. Rev.* **105** (1957), 508.
[||] H. Haken. *Fortschritte der Physik*, **6** (1958), 271.
[¶] For details concerning the band structure of these two materials, consult F. Herman, *Proc. Inst. Radio Engrs*, *N.Y.*, **43** (1955), 1703.

the split-off point; this state has a $\mathbf{k}_0$ different from zero, which means that the wave function has a substantial phase shift from cell to cell. Such a variation brings us into conflict with condition (b) of p. 177, and thus does not allow us to expand $W(\mathbf{k})$ directly, as in (6.25). We must write out this rapid variation explicitly first; that is, use the function

$$\psi_q(\mathbf{r}) = e^{-2\pi i \mathbf{k}_0 \cdot \mathbf{r}} \phi_q(\mathbf{r}) \tag{6.43}$$

as our slowly varying wave function. With this new amplitude function the wave function of the excess carrier takes the form

$$\Psi(\mathbf{x}) = \sum_{\mathbf{r}} \psi_q(\mathbf{r}) \operatorname{ixp} \mathbf{k}_0 \cdot \mathbf{r} \; a(\mathbf{x} - \mathbf{r}), \tag{6.44}$$

and the operator equation (6.30) for $\phi_q(\mathbf{r})$ becomes

$$\{\operatorname{ixp}(-\mathbf{k}_0 \cdot \mathbf{r}) \, W_q(\mathbf{k}) \operatorname{ixp}(\mathbf{k}_0 \cdot \mathbf{r}) + V(\mathbf{r})\} \psi_q(\mathbf{r}) = E \psi_q(\mathbf{r}). \tag{6.45}$$

The reader can easily verify from (6.19) that the following identity holds

$$\operatorname{ixp}(-\mathbf{k}_0 \cdot \mathbf{r}) \, F(\mathbf{k}) \operatorname{ixp} \mathbf{k}_0 \cdot \mathbf{r} = F(\mathbf{k} + \mathbf{k}_0). \tag{6.46}$$

With the help of this identity and the equations (5.52) and (6.21), the relation (6.45) can be given the form

$$\left\{ W_q\left(\mathbf{k}_0 - \frac{i}{2\pi} \frac{\partial}{\partial \mathbf{r}}\right) + V(\mathbf{r}) \right\} \psi_q(\mathbf{r}) = E \psi_q(\mathbf{r}). \tag{6.47}$$

In other words, as we write $\phi_q(\mathbf{r})$ in the form

$$\phi_q(\mathbf{r}) = \operatorname{ixp} \mathbf{k}_0 \cdot \mathbf{r} \; \psi_q(\mathbf{r}), \tag{6.48}$$

with $\psi_q(\mathbf{r})$ as our unknown slowly varying function, and as we satisfy thereby condition (b) of p. 177, we find that we have to develop now $W_q(\mathbf{k})$ around $\mathbf{k}_0$ as the new starting point for an expansion of the form (6.25). Around this new point the development may proceed as in (5.60), yielding

$$\left\{ -\frac{1}{8\pi^2} \sum_{i,j=1}^{3} \alpha_{ij} \frac{\partial^2}{\partial r_i \partial r_j} + V(\mathbf{r}) + W(\mathbf{k}_0) - E \right\} \psi_q(\mathbf{r}) = 0. \tag{6.49}$$

The equation can be simplified slightly beyond this point by choosing co-ordinates $i = 1, 2, 3$ in such a way that the tensor of the α_{ij} is reduced to principal axes. We then deal with three coefficients which are usually written in the form

$$\alpha_{ii} = h^2/m_i, \tag{6.50}$$

that is, we have three effective masses. Further simplification of the wave equation (6.49) is not possible, because it must reflect in an essential way the fact that the band energy is related to the crystal momentum by a tensor of effective reciprocal mass.

The next stage of complication arises for defect electrons in germanium. The energy maximum of the valence band, which is their energy minimum, is a degenerate state at $\mathbf{k} = 0$. Consequently, the energy is not developable in power series at all, and condition (a) of p. 177 is not fulfilled. In order to circumvent this difficulty we have to revert to the not fully diagonalized Bloch functions $\beta_q^\nu(\mathbf{x}; \mathbf{k})$ of equation (5.54). It is seen from that equation that the result of operating on these functions with the crystal hamiltonian is a finite matrix of elements $w_q^{\mu\nu}(\mathbf{k})$; the energy eigenvalues are obtained from that matrix by annulling the secular determinant (5.55). The elements $w_q^{\mu\nu}(\mathbf{k})$ have the advantage of being developable in powers of $\mathbf{k}$. Hence we may use these functions $\beta_q^\nu(\mathbf{x}; \mathbf{k})$ as a starting base for the unperturbed hamiltonian, writing

$$\langle g, q, \nu; \mathbf{k} | \mathcal{H}_0 | g, q', \nu'; \mathbf{k}' \rangle = \delta_{q,q'} \delta_{\mathbf{k},\mathbf{k}'} w_q^{\nu\nu'}(\mathbf{k}). \tag{6.51}$$

If we now transform from these Bloch functions to cellular functions $a_q^\nu(\mathbf{x} - \mathbf{r})$ by the transformation (6.15), and define the operator $\mathbf{r}$ by (6.18) as previously, then relation (6.19) will follow, with an extra factor $\delta_{\nu,\nu'}$ on the right. Hence the symbolic operator identities (6.20) and (6.21) will hold as previously. We end up, therefore, with a matrix relation of the form

$$\langle q, \nu; \mathbf{r} | \mathcal{H}_0 | q', \nu'; \mathbf{r}' \rangle = \delta(\mathbf{r} - \mathbf{r}') \delta_{q,q'} w_q^{\nu\nu'} \left(-\frac{i}{2\pi} \frac{\partial}{\partial \mathbf{r}} \right). \tag{6.52}$$

This new form must now replace the previously used expression (6.6) in (6.27). The argument subsequent to (6.27) is not affected by the new form of $\mathcal{H}_0$, and hence the substitution (6.52) for $W(\mathbf{k})$ may also be made in equation (6.30). This new form of (6.30) is now legitimately approximated by the differential expressions (6.25), This means that we are now dealing with a system of coupled differential equations. Solution of such a system is obviously much more difficult than of a single equation; it will not be discussed here, and those who want further information are referred to the literature.†

DYNAMICS OF ELECTRONS IN ELECTRIC AND MAGNETIC FIELDS

Having discussed the result of a hamiltonian of the form (6.30) for the cases in which it leads to a small number of quantized states, we want to proceed to other situations. These situations lead either to a continuous spectrum or to quantized states of very high quantum number. General experience indicates that, for these cases, classical mechanics represents a fair approximation. In a continuum the equivalence of the two types of mechanics is assured by the possibility of forming wave packets obeying classical laws; in elevated quantum states the equivalence is furnished by Bohr's correspondence principle which identifies classical periods with quantum energy spacings. The limitation to the analogy is often furnished by the tunnel effect. Because of this effect certain processes occur in quantum theory with small probability, while classically they should not occur at all. These processes, even though improbable, can be important on physical grounds, particularly if they bring about the decay of a classically stable system; radioactivity is the model case for this situation.

Reversion to classical mechanics for electrons in solids is not an obvious procedure, and has been delayed much too long. The reason is that the reversion cannot be a *direct* reversion. The important energy bands in solid state theory are low quantum number bands. They arise quite essentially through the wave character of the electron; no classical procedure could hope to predict their properties. In fact, quantum mechanics itself has difficulties in doing so. However, within the bands we deal with a quasicontinuum containing an enormous number of states; from these states wave packets can be formed. It follows from equation (6.15) that these wave packets can be shrunk essentially to the dimension of one primitive cell. Therefore, if a phenomenon covers many primitive cells then we should not hesitate in employing classical procedures, keeping in mind of course their inherent limitations.

† W. Kohn and D. Schechter, *Phys. Rev.* **99** (1955), 1903.

The natural variables for such a quasiclassical treatment are the *crystal momentum* $h\mathbf{k}$ and the *lattice vector* $\mathbf{r}$. It follows from (6.19), (6.20) and (6.21) that they are conjugate and commute with q. If we leave aside problems arising from band degeneracy then our hamiltonian for fixed q is given by (6.30):

$$\mathcal{H} = W_q(\mathbf{k}) + V(\mathbf{r}). \qquad (6.53)$$

Hamilton's equations derived from this read

$$h\frac{d\mathbf{r}}{dt} = \frac{\partial W_q(\mathbf{k})}{\partial \mathbf{k}}, \qquad (6.54)$$

$$h\frac{d\mathbf{k}}{dt} = -\frac{\partial V(\mathbf{r})}{\partial \mathbf{r}}. \qquad (6.55)$$

Both (6.54) and (6.55) repeat known results if $V(\mathbf{r})$ equals a constant. (6.54) repeats (6.24), (6.55) repeats that $\mathbf{k}$ is a good quantum number. New results arise when $V(\mathbf{r})$ is not constant. Let us take for closer study the case of a *uniform electric field*:

$$V(\mathbf{r}) = -e\,\mathbf{E}\cdot\mathbf{r}. \qquad (6.56)$$

For this case equation (6.55) takes the form

$$h\frac{d\mathbf{k}}{dt} = e\,\mathbf{E}. \qquad (6.57)$$

Integration of this yields, with $\mathbf{E}$ along the x-direction

$$k_x - \frac{e}{h}Et = k_0 = \text{constant}, \qquad (6.58a)$$

$$k_y = \text{constant}, \qquad (6.58b)$$

$$k_z = \text{constant}. \qquad (6.58c)$$

(6.58a) is a generalization to periodic hamiltonians of the law of constant acceleration in uniform fields; the other two equations contain nothing new. Among the equations (6.54), only the x-equation gives an interesting result

$$r_x(t) = r_x(0) + \frac{1}{h}\int_0^t \frac{\partial W_q(\mathbf{k})}{\partial k_x}\,dt$$

$$= r_x(0) + \frac{1}{eE}\int_{k_0}^{k_x} \frac{\partial W_q(\mathbf{k})}{\partial k_x}\,dk_x,$$

or $\qquad r_x(t) = r_x(0) + \frac{1}{eE}\{W_q(k_x, k_y, k_z) - W_q(k_0, k_y, k_z)\}. \qquad (6.59)$

For simplicity of discussion let $\mathbf{E}$ be perpendicular to a lattice plane of the crystal; the y-z plane is then this lattice plane, and by the theorem (a) of Chapter 2, the x-direction is a direction of the reciprocal lattice. Let the period in that direction be a^*. Then the motion in $\mathbf{k}$-space is periodic in time; for from (6.58), the carrier reaches an equivalent point after a period T given by

$$T = \frac{ha^*}{eE}. \tag{6.60}$$

We can give this another form by using theorem (b) of Chapter 2 according to which a^* is in magnitude the reciprocal distance of successive y-z lattice planes. Now let the crystal lattice be described by two vectors in the y-z plane and a vector $\mathbf{a}$ out of the plane (usually not parallel to $\mathbf{E}$). Then we have

$$a^* = \frac{1}{a\cos\left(\mathbf{a},\mathbf{E}\right)},$$

and hence

$$T = \frac{h}{e\left(\mathbf{E}\cdot\mathbf{a}\right)}. \tag{6.61}$$

Perusal of (6.59) shows that the motion is not only periodic in $\mathbf{k}$-space, but also along the x-direction since

$$r_x(T) = r_x(0).$$

A drift in the y-z plane, that is at right angles to the field, can generally take place, in which equal distances are traversed in equal periods T. Finally, it goes without saying that this analysis is completely upset if many collisions occur in a period T. This is the normal situation, as the period between collisions is usually of the order 10^{-12} sec., and rather large fields are required to keep T from being larger than this number.

    The conclusions obtained here are sufficiently startling to justify a look at the quantum situation. The two conclusions to verify are that $k_x - (eEt/h)$ is a constant of the motion as exhibited in (6.58), and that, except for a drift at right angles to the field, the motion is periodic with a period T given by (6.61). If we take for the hamiltonian (6.53) and (6.56) both conclusions are easily verified. Writing

$$\mathcal{H} = W_q(\mathbf{k}) - e\,\mathbf{E}\cdot\mathbf{r} \tag{6.62}$$

we get
$$i\hbar \frac{d}{dt}\left(k_x - \frac{eEt}{h}\right) = i\hbar \frac{\partial}{\partial t}\left(k_x - \frac{eEt}{h}\right)$$

$$-\left[\mathscr{H}\left(k_x - \frac{eEt}{h}\right) - \left(k_x - \frac{eEt}{h}\right)\mathscr{H}\right]$$

$$= -\frac{ieE}{2\pi} + eE(r_x k_x - k_x r_x)$$

$$= 0.$$

from (6.20). The existence of a period T given by (6.61) expresses itself in quantum theory as a Stark splitting within each band of width h/T or $e\mathbf{E}\cdot\mathbf{a}$. This is easily verified from (6.62). Suppose we have an amplitude function $\phi_q(\mathbf{r})$ solving (6.62),

$$\{W_q(\mathbf{k}) - e\mathbf{E}\cdot\mathbf{r}\}\phi_q(\mathbf{r}) = \mathscr{E}\phi_q(\mathbf{r}),$$

then the geometrically displaced function $\phi_q(\mathbf{r}+\mathbf{a})$ obeys the relation

$$\{W_q(\mathbf{k}) - e\mathbf{E}\cdot(\mathbf{r}+\mathbf{a})\}\phi_q(\mathbf{r}+\mathbf{a}) = \mathscr{E}\phi_q(\mathbf{r}+\mathbf{a}).$$

This means, however, that $\phi_q(\mathbf{r}+\mathbf{a})$ is a solution of the same hamiltonian with shifted energy, namely

$$\{W_q(\mathbf{k}) - e\mathbf{E}\cdot\mathbf{r}\}\phi_q(\mathbf{r}+\mathbf{a}) = (\mathscr{E} + e\mathbf{E}\cdot\mathbf{a})\phi_q(\mathbf{r}+\mathbf{a}),$$

which is the sought-for Stark splitting. The reason these quantum proofs are not satisfactory is that the true hamiltonian is not (6.62), but

$$\mathscr{H} = W_q(\mathbf{k}) - e\mathbf{E}\cdot\mathbf{x}. \tag{6.63}$$

Here the last term does not commute with q and this produces interband transitions. In spite of this, it is still possible to prove the relation (6.58), the 'law of constant acceleration'. Furthermore the proof of a Stark splitting of the type

$$\Delta\mathscr{E} = e\mathbf{E}\cdot\mathbf{a} \tag{6.64}$$

within each band holds also because the proof given transfers to the hamiltonian (6.63). However, electrons can now escape strict periodicity in $\mathbf{k}$-space by transitions from band to band through the intervening forbidden gap. This tunnelling process is known as *Zener effect*.†

† C. Zener, *Proc. Roy. Soc. A*, **145** (1934), 523. W. V. Houston, *Phys. Rev.* **57** (1940), 184.

The best starting point for a theory of the action of a *magnetic field* on electrons is to take the standpoint of special relativity, according to which there is no separate magnetic force on a charge, but the electron is simply acted upon by the electric force acting in its own frame of reference. If this starting point is accepted then the action of a magnetic field upon a charge is determined by the appropriate Lorentz transformation. The result is that the electric and magnetic field act in a package, the so-called *Lorentz force* **F** which is given by †

$$\mathbf{F} = \mathbf{E} + \frac{1}{c}\mathbf{v} \times \mathbf{H}. \tag{6.65}$$

If this is correct, and if our equations of motion in the absence of a magnetic field are (6.54) and (6.57), then their logical extension is

$$h\frac{d\mathbf{r}}{dt} = \frac{\partial W_q(\mathbf{k})}{\partial \mathbf{k}}, \tag{6.66}$$

$$h\frac{d\mathbf{k}}{dt} = e\left(\mathbf{E} + \frac{1}{c}\frac{d\mathbf{r}}{dt} \times \mathbf{H}\right). \tag{6.67}$$

Errors similar to the 'loss' of the Zener effect must be expected from these equations; for it is seen from (5.59) and (6.24) that the velocity and $d\mathbf{r}/dt$ have equal diagonal elements; the velocity operator has, however, off-diagonal elements in addition, which will produce interband effects analogous to the Zener effect.

The most direct applications of the Lorentz force law is the theoretical interpretation of the *Hall effect*. To do this we accept without further discussion Ohm's law (discussed in the next chapter), according to which real electrons in real solids are subject to a frictional force which is such that they acquire a drift velocity proportional to the applied Lorentz force; in cubic materials this law makes the drift velocity also parallel to the force. Since the force is given by (6.65), Ohm's law in the presence of a magnetic field must read

$$\mathbf{i} = \sigma\mathbf{F}, \tag{6.68}$$

where **i** is the current density, σ the conductivity, and **F** is given by (6.65). Furthermore, if the carriers are only of one kind, and if their movement is sufficiently similar that averaging over them has

† See for instance R. Becker, *Theorie der Elektrizität* (Teubner, 1933), Vol. 2, § 56.

a meaning, then the current is related to their mean drift velocity **v** by the formula

$$\mathbf{i} = n e \mathbf{v}, \tag{6.69}$$

where n is the carrier density. This drift velocity **v**, being the average velocity, is also the one entering into (6.65). If one now observes the flow of current in a specimen under the influence of crossed electric and magnetic fields, then the current will be perpendicular to the magnetic field because **F** is perpendicular to it by (6.65). It will not be parallel to **E**, however, but make with it

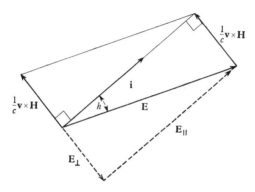

Fig. 6.4. Vectors and angles in the theory of the Hall effect. The magnetic field **H** is assumed perpendicular to the paper; the Hall angle h is defined as the angle between the electric field **E** and the current density **i**.

a *Hall angle h*. In fig. 6.4 are drawn the vectors **i**, **E** and $(1/c)\,\mathbf{v} \times \mathbf{H}$, with **H** assumed perpendicular to the paper. One reads immediately off the drawing that

$$\sin h = \frac{1}{c} v \frac{H}{E}, \tag{6.70}$$

that is, the drift velocity is directly determinable from the fields and the Hall angle. Another version of the same interpretation is often used, which eliminates the electric field parallel to the current. If the electric field is decomposed into a component perpendicular and parallel to **i** as shown by the dashed vectors in fig. 6.4 then $E_\perp$ must equal the magnetic part of the Lorenz force, viz.

$$E_\perp = \frac{1}{c} v H;$$

Division by iH and utilization of (6.69) leaves the customary form

$$\frac{E_\perp}{iH} = \frac{1}{nec}. \tag{6.71}$$

The left-hand side is the definition of the Hall constant, and the right-hand side gives its interpretation, in the circumstances stated, in terms of the carrier density n. The use of (6.71) for the measurement of carrier density is widespread in semiconductor work. Unfortunately, the formula is often invalidated by the presence of two different types of carriers, such as excess and defect electrons. The fact that the carriers behave more like a gas than a liquid, and have therefore a maxwellian kind of velocity distribution, leads only to an insignificant correction factor in the formula.

Interest has shifted in recent years from the purely frictional type of electron motion to electron dynamics in the absence of friction. The frictional aspect of electronic motion is of the collision type, and can be minimized by observations at high frequency. Observations of the Zeemann pattern into which the levels of a band break up in a magnetic field appears particularly promising. The experiment consists in inducing transitions among these levels by a microwave field. When the microwave frequency matches the Zeemann splitting a resonance phenomenon is observed. The phenomenon is known as *cyclotron resonance* or *diamagnetic resonance*. The dynamical problem can be sketched, as usual, without explicit consideration of the microwave field; it is simply the problem of the motion of an electron in a uniform magnetic field. We shall treat this problem first with the quasi-classical equations (6.66) and (6.67); the splitting will then appear as the period of a cyclical classical motion, as explained earlier in connection with the Stark splitting of bands by an electric field. This type of calculation was first performed by Landau[†] for quasifree electrons and by Shockley[‡] for electrons in arbitrary bands.

A glance at the equations (6.66) and (6.67) shows that it is advantageous to think of the motion of the electrons in **k**-space first

---

† L. Landau. *Zeits. f. Phys.* **64** (1930), 629.

‡ W. Shockley, *Phys. Rev.* **79** (1950), 191. It is not clear from the brief article that the quantization rules for an electron in a magnetic field are in fact derived. For a fuller discussion of the implications see I. M. Lifshitz and A. M. Kosevitch, *JETP* **29** (1955), 730 [*Sov. Phys. JETP* **2** (1956), 636].

and treat the motion in r-space subsequently; for elimination of dr/dt between them leaves an equation in the three components of $\mathbf{k}$ alone and thus represents a reduction to a smaller number of variables. Let us write out the three resulting equations, assuming the electric field zero, and the magnetic field along the z-direction. We get

$$\frac{dk_x}{dt} = \frac{eH}{h^2c}\frac{\partial W_q(\mathbf{k})}{\partial k_y}, \qquad (6.72a)$$

$$\frac{dk_y}{dt} = -\frac{eH}{h^2c}\frac{\partial W_q(\mathbf{k})}{\partial k_x}, \qquad (6.72b)$$

$$\frac{dk_z}{dt} = 0. \qquad (6.73)$$

Rather surprisingly, the equations (6.72) are in themselves a hamiltonian system of relations for one degree of freedom with $(h^2c/eH)^{\frac{1}{2}}k_x$ and $(h^2c/eH)^{\frac{1}{2}}k_y$ playing the role of conjugate variables; according to (6.73), k_z acts as a fixed equation parameter of no dynamical status. The trajectories lie in the k_x-k_y plane; they have an energy integral, that is, it follows from (6.72) that

$$\frac{dW_q(\mathbf{k})}{dt} = \frac{\partial W_q(\mathbf{k})}{\partial k_x}\frac{dk_x}{dt} + \frac{\partial W_q(\mathbf{k})}{\partial k_y}\frac{dk_y}{dt} = 0. \qquad (6.74)$$

The trajectories are therefore the constant energy curves which are the intersections of the constant energy surfaces in $\mathbf{k}$-space with planes $k_z = $ constant. These curves are either open, in which case they are periodic in reciprocal space (because equivalent points in reciprocal space have equal energy), or they are closed. In either case the motion in $\mathbf{k}$-space is periodic in time. Specializing for the latter case, we may calculate the period T for traversing the closed curve by the formula

$$\begin{aligned} T &= \oint \frac{dk_x}{\dot{k}_x} = \frac{h^2c}{eH}\oint\frac{dk_x}{(\partial W/\partial k_y)_{k_x,k_z}}\\ &= \frac{h^2c}{eH}\oint\left(\frac{\partial k_y}{\partial W}\right)_{k_x,k_z}dk_x\\ &= \frac{h^2c}{eH}\left\{\frac{\partial}{\partial W}\oint k_y\,dk_x\right\}_k. \end{aligned}$$

The integral is the area A enclosed by the trajectory having the energy W in a plane $k_z = $ constant. Hence we can write for the period T

$$T = \frac{h^2 c}{eH}\left(\frac{\partial A}{\partial W}\right)_{k_z}. \tag{6.75}$$

As the formula is to be applied to semiconductor carriers we may consider ourselves at a non-degenerate extremum and apply the simplification (5.60). Let us diagonalize the mass tensor as in (6.50) to give $W(\mathbf{k})$ the form

$$W(\mathbf{k}) = \frac{h^2}{2}\left[\frac{(k_x - k_{x0})^2}{m_x} + \frac{(k_y - k_{y0})^2}{m_y} + \frac{(k_z - k_{z0})^2}{m_z}\right]. \tag{6.76}$$

Let the magnetic field be along direction cosines α, β, γ. We then have to hold constant the component of $\mathbf{k}$ along that direction, that is

$$\alpha(k_x - k_{x0}) + \beta(k_y - k_{y0}) + \gamma(k_z - k_{z0}) = k \tag{6.77}$$

must be a constant. This means that we must calculate the area of the intersection of the plane (6.77) and the ellipsoid (6.76) and then insert its derivative with respect to W into (6.75). The easiest way to compute that area A is to compute the volume Ω of the cone which it subtends with the point $\mathbf{k}_0$ as apex. We can then use the volume formula for the cone to compute A:

$$A = \frac{3}{k}\Omega. \tag{6.78}$$

In this way affine transformations can be used freely. The transformation

$$\left.\begin{aligned} x &= \frac{k_x - k_{x0}}{\sqrt{m_x}}, \\ y &= \frac{k_y - k_{y0}}{\sqrt{m_y}}, \\ z &= \frac{k_z - k_{z0}}{\sqrt{m_z}}, \end{aligned}\right\} \tag{6.79}$$

transforms the ellipsoid (6.76) into the sphere

$$W = \frac{h^2}{2}(x^2 + y^2 + z^2). \tag{6.80}$$

The plane (6.77) becomes

$$\alpha m_x^{\frac{1}{2}} x + \beta m_y^{\frac{1}{2}} y + \gamma m_z^{\frac{1}{2}} z = k. \tag{6.81}$$

The cone discussed on the preceding page is now a circular cone of volume Ω_c. It follows from (6.79) that Ω and Ω_c are related through

$$\Omega = (m_x m_y m_z)^{\frac{1}{2}} \Omega_c,$$

and therefore with (6.78)

$$A = \frac{3}{k}(m_x m_y m_z)^{\frac{1}{2}} \Omega_c. \tag{6.82}$$

Computation of the volume Ω_c of the circular cone subtended by the intersection of (6.80) and (6.81) at the origin is now elementary. We find

$$\Omega_c = \tfrac{1}{3}\pi \left(\frac{2W}{h^2} - \frac{k^2}{\alpha^2 m_x + \beta^2 m_y + \gamma^2 m_z}\right) \frac{k}{(\alpha^2 m_x + \beta^2 m_y + \gamma^2 m_z)^{\frac{1}{2}}}.$$

Insertion of this yields finally

$$A = \frac{2\pi}{h^2} \frac{1}{\left(\dfrac{\alpha^2}{m_y m_z} + \dfrac{\beta^2}{m_z m_x} + \dfrac{\gamma^2}{m_x m_y}\right)^{\frac{1}{2}}} \left\{ W - \frac{1}{2} \frac{h^2 k^2}{\alpha^2 m_x + \beta^2 m_y + \gamma^2 m_z}\right\},$$

$$\tag{6.83}$$

from which we get by (6.75)

$$T = \frac{2\pi c}{eH} \frac{1}{\left(\dfrac{\alpha^2}{m_y m_z} + \dfrac{\beta^2}{m_z m_x} + \dfrac{\gamma^2}{m_x m_y}\right)^{\frac{1}{2}}}. \tag{6.84a}$$

This is equivalent to a mean angular velocity ω which equals

$$\omega = \frac{e}{c} H \left(\frac{\alpha^2}{m_y m_z} + \frac{\beta^2}{m_z m_x} + \frac{\gamma^2}{m_x m_y}\right)^{\frac{1}{2}}. \tag{6.84b}$$

In particular, when the simple situation (5.61) prevails, the frequency becomes independent of the relative orientation of field and crystal and equals

$$\omega = \frac{e}{m^* c} H. \tag{6.85}$$

The formal combination of constants entering into (6.85) is the same as the operating frequency of a cyclotron. It differs from the Larmor frequency controlling the Zeemann splitting of bound levels by a factor 2. Hence the name of cyclotron resonance. The usefulness of (6.84) and (6.85) for the determination of effective masses of semiconductor electrons is obvious.

For a quantum theoretical treatment of cyclotron resonance the Hamiltonian form of the equations (6.66) and (6.67) must first be restored. This form was destroyed by the introduction of magnetic effects through the Lorentz force (6.65). The restoration of the Hamiltonian form is achieved by retaining $\mathbf{r}$ as a variable and retaining the hamiltonian (6.53), but replacing $\mathbf{k}$ by a new momentum variable which is defined through

$$\mathbf{k} = \frac{1}{h}\left(\mathbf{p} - \frac{e}{c}\mathbf{A}(\mathbf{r})\right). \qquad (6.86)$$

Indeed, Hamilton's equations derived from the hamiltonian

$$\mathscr{H} = W_q\left(\frac{1}{h}\mathbf{p} - \frac{e}{hc}\mathbf{A}(\mathbf{r})\right) + eV(\mathbf{r}) \qquad (6.87)$$

reproduce the equations (6.66) and (6.67) if the usual equations for the fields in terms of the potentials

$$\mathbf{H} = \nabla \times \mathbf{A},$$

$$\mathbf{E} = -\nabla V - \frac{1}{c}\frac{\partial \mathbf{A}}{\partial t},$$

are employed in the derivation; the interpretation must be adopted that the operator ∇ differentiates with respect to the lattice vector $\mathbf{r}$. The reader is invited to carry through this argument in detail. The hamiltonian (6.87) opens an extremely attractive way of applying quantum mechanics to the problem of computing the Zeemann splitting of energy bands by a magnetic field. We can reintroduce the components of $\mathbf{k}$ by (6.86), just as we did in the classical calculation of the effect. These components are now no longer the conjugate variables of $\mathbf{r}$; in fact, they no longer commute with each other. The basic conjugate variables are $\mathbf{p}$ and $\mathbf{r}$, and (6.86) must be employed to find the commutation relations of the components of $\mathbf{k}$ among themselves. We find

$$\mathbf{k} \times \mathbf{k} = -\frac{ie\mathbf{H}}{2\pi hc}. \qquad (6.88)$$

If $\mathbf{H}$ is along the z direction, k_z commutes with everything else, and $(h^2c/eH)^{\frac{1}{2}}k_x$ and $(h^2c/eH)^{\frac{1}{2}}k_y$ become conjugate variables.

This feature resulted already from the preceding semi-classical calculation and was pointed out in connexion with the equations (6.72). If there is no $V(\mathbf{r})$ the hamiltonian (6.87) simply takes the form

$$\mathcal{H} = W_q(\mathbf{k}), \tag{6.89}$$

that is, it is formally the same as without a magnetic field. The magnetic field is then brought in through the anomalous commutation rules (6.88). The combination of (6.88) and (6.89) is now a self-contained quantum mechanical problem for the determination of the Zeemann levels within a band. Solution of this problem is found to confirm (6.84) exactly, and the general formula (6.75) in the Bohr–Sommerfeld limit of quantum mechanics; this means that small deviations from (6.75) must be expected for low quantum numbers, that is at the band edges. These two calculations will be left as exercises in quantum mechanics for the reader.

It is unfortunate that the argument for the hamiltonian (6.87). presented here is not conclusive from a quantum viewpoint, because of its reliance on classical hamiltonian transformation theory. In our present understanding of quantum mechanics, cartesian co-ordinates have a privileged position: the quantum hamiltonian is to be defined as the cartesian hamiltonian, plus the commutation relations of cartesian co-ordinates and momenta. Because of these commutation relations, the classical contact transformations to other variables can be duplicated in quantum theory only with the appearance of extra terms arising from the commutators. The hamiltonian (6.87) is therefore only made plausible as a quantum hamiltonian by the derivation given here. Kohn† has given a derivation which is not subject to this criticism.

PROBLEMS

1. Prove that the Fermi–Dirac 'distribution law (5.62) entails for excess and defect electrons a mass action law; the product of their densities equals a constant dependent on temperature, but not on the position of the Fermi level; the statement holds as long as neither one of the carriers is degenerate.

† W. Kohn, *Phys. Rev.* **115** (1959), 1460.

2. Show that if the approximation (6.50) is valid for the minimum of the conduction band and the maximum of the valence band then the constant of problem 1 equals

$$\frac{32\pi^3(kT)^3}{h^6}(m_{n1}m_{n2}m_{n3}m_{p1}m_{p2}m_{p3})^{\frac{1}{2}}\exp\left(-\frac{\text{band gap}}{kT}\right),$$

where m_{ni} and m_{pi} are the three effective masses of the negative and positive carriers respectively. Show also that if there are several minima or maxima the product of the number of each must be added as a factor.

3. Show that if the Born–von Karman conditions are dropped and the crystal volume is considered infinite, Bloch functions can be normalized either by normalization within a primitive cell or by a condition of the form

$$\int_\Omega b_q^*(\mathbf{x};\mathbf{k})\,b_q(\mathbf{x};\mathbf{k}')\,d\mathbf{x} = \delta(\mathbf{k}-\mathbf{k}').$$

Find the relationship between the two normalizations.

4. Show that the cellular functions defined by (6.15) remain normalizable and orthogonal under the conditions of problem 3. Give their proper defining equation which is to replace (6.15) under the two situations envisaged in problem 3.

5. Prove the identity (6.46) of the text.

6. Instead of the form (6.44) of the wave function Ψ, Luttinger and Kohn† have proposed to write the wave function in the form of a product

$$\Psi(\mathbf{x}) = \psi_q(\mathbf{x})\,b_q(\mathbf{x};\mathbf{k}_0),$$

where $\psi_q(z)$ is the same function as in the text of this chapter. Show that this form is approximately correct, the error being

$$\frac{\partial\psi}{\partial\mathbf{x}}\cdot\left(\mathbf{x}+\frac{i}{2\pi}\frac{\partial}{\partial\mathbf{k}_0}\right)b(\mathbf{x};\mathbf{k}_0).$$

7. Suppose we have a band in two dimensions whose energy as a function of $\mathbf{k}$ is given by

$$W = W_0\cos 2\pi k_x d \cos 2\pi k_y d.$$

Determine the Koster-Slater level E_0 by the method developed in the text, and show that it is given by the relation

$$V_0 = \frac{\pi E_0}{2K(W_0/|E_0|)}$$

where K is the elliptic integral of the first kind. In what respect does this result deviate most markedly from the result exhibited in fig. 6.3?

8. Suppose we have a crystal having a body-centred cubic Bravais lattice and suppose the energy $W_q(\mathbf{k})$ of an allowed band has the form

$$W_q(\mathbf{k}) = 1 - \cos 2\pi k_x a \cos 2\pi k_y a \cos 2\pi k_z a,$$

where a, the elementary cube edge, equals 2 Å, and $W_q(\mathbf{k})$ is measured in electron volts. Suppose now that all other bands are far removed from

† J. M. Luttinger and W. Kohn, *Phys. Rev.* **97** (1955), 869.

this one and suppose an impurity state for an electron is found 1/10 e.v. below the band. What information can you derive from these data for the behaviour of the wave function of an electron in this state if you assume the wave function to have cubic symmetry?

9. Supply a proof for the statement made in the text that the conservation law (6.58) for $\mathbf{k} - e/h\ \mathbf{E}t$ is exact, even if quantum mechanics and the exact hamiltonian (6.63) are employed. (*Note*: The problem is much easier in the so-called time gauge for the electric field, $V = 0$, $\mathbf{A} = \mathbf{E}ct$).

10. When a piece of germanium uniformly doped with donors is in contact with a similar piece doped with acceptors, an n-p junction develops as explained in the text. In the course of this adjustment the layer immediately adjoining the junction is drained of electric carriers and becomes an insulator (exhaustion layer). In order to estimate the width of this layer, assume the p side much more heavily doped than the n side. and assume kT/e small compared to the shift V_0 of the Fermi level between the two sides. Show that under these conditions the exhaustion layer is entirely on the n side, and that its thickness t equals

$$t = \left(\frac{V + V_0}{2\pi eN}\right)^{\frac{1}{2}},$$

where V is the voltage applied across the junction and N the density of donor centres. Put representative numbers in for N (see beginning of chapter), V and V_0, to get the range of variation of t.

11. Show that

$$\mathscr{H} = \frac{1}{2m}\left(\mathbf{p} - \frac{e}{c}\mathbf{A}(\mathbf{x})\right)^2 + eV(\mathbf{x})$$

is a correct hamiltonian for a classical particle of charge e, moving in an electromagnetic field. Go on to prove that (6.66) and (6.67) result from a band hamiltonian of the type (6.87).

12. Solve exactly the Schrödinger equation for a free electron in a uniform magnetic field along the z-direction using the hamiltonian derived in the preceding problem. Solve it first in cartesian co-ordinates with the gauge

$$A_x = 0, \quad A_y = Hx, \quad A_z = 0$$

and then in cylindrical co-ordinates with the gauge

$$A_x = -\tfrac{1}{2}Hy, \quad A_y = \tfrac{1}{2}Hx, \quad A_z = 0.$$

Show that the energy states are discrete for each fixed value of p_z and that the splitting is in accord with the cyclotron formula (6.85). Comment on the discrepancy of the wave functions in the two derivations.

13. Calculate to first order the splitting induced by a magnetic field in a discrete (other than s) state of an electron in a spherically symmetrical field. Show that the splitting is half that of problem 12 (Larmor frequency).

14. Show by quantum mechanical reasoning the correctness of (6.75) for high quantum numbers, using (6.88) and (6.89) as your starting point

and employing the Bohr–Sommerfeld quantum conditions. The differentials in (6.75) must in this derivation be considered equivalent to the small differences between the quantities in neighbouring discrete states.

15. Prove in two steps that the result (6.84) is *exact* in quantum mechanics. First prove it for the case that the three effective masses are equal, and thereupon, employ the transformation (6.79) to reduce the general case to the one treated initially.

BIBLIOGRAPHY

GENERAL TREATISES

N. B. HANNAY et al. *Semiconductors*. (New York, Reinhold, 1959.)

O. MADELUNG. Halbleiter. *Handbuch der Physik*, **20** (1957), 1.

W. SHOCKLEY, *Electrons and Holes in Semiconductors with Applications to Transistor Electronics*. (New York, Van Nostrand, 1950.)

E. SPENKE. *Elektronische Halbleiter*. (Berlin, Springer, 1956.)

WORKS SPECIALIZED TO GERMANIUM AND SILICON

H. Y. FAN. Valence semiconductors, germanium and silicon. *Solid State Physics*, **1** (1955), 284–365.

Transistor issue. *Proc. Inst. Radio Engrs.*, *N.Y.*, **40** (1952), 1283–1602.

F. HERMAN. The electronic energy band structure of silicon and germanium. *Proc. Inst. Radio Engrs. N.Y.*, **43** (1955), 1703–32.

W. KOHN. Shallow impurity states in silicon and germanium. *Solid State Physics*, **5** (1957), 257–320.

J. N. SHIVE. *Properties, Physics and Design of Semiconductor Devices*. (New York, Van Nostrand, 1959.)

OTHER SEMICONDUCTOR TYPES

H. WELKER and H. WEISS. Group III–Group V Compounds. *Solid State Physics*, **3** (1956), 1–78.

CHAPTER 7

FRICTION IN ELECTRON DYNAMICS; OHM'S LAW

The device of ignoring friction is widespread in all branches of mechanics. It is particularly useful when qualitative insight is desired in an early superficial study. Later on one may wish to throw a second glance upon the subject. One often finds at this stage that energy dissipation can simply be grafted upon the previously developed conservative formalism. Finally, a third stage can be envisaged, in which it is shown how friction can arise out of the general laws of nature which are believed to conserve energy.

There are many gaps in our knowledge of the third stage, even in large scale mechanics. In quantum mechanics the recognition of friction as a distinct ingredient is relatively recent, and all that can be said at present is that it has been proved in certain situations not to be in contradiction with the fundamental laws of mechanics.[†] However, if one is willing to grant the premise that a certain perturbing interaction leads to relaxation processes rather than to a recognizably different set of stationary states, then the computation of the relaxation time and the grafting process indicated can be extended with relative ease from classical to quantum mechanics. This section will be exclusively concerned with the application of this second stage to quasi-free electrons in solids. The result is a theory which has a very much broader contact with experimental reality.

The central theme for this sort of study is *Ohm's law*. Ohm's law states that electrons in solids acquire a constant drift velocity under the influence of a constant electric field. Such a behaviour is in contradiction, not only to the theory of electronic motion developed in the preceding chapter, but also to the easily verified laws of motion of electrons in a vacuum. Clearly, a solid is not vacuum for the electrons within it, but produces frictional drag upon them which is distinct from, and superimposed upon, the

† L. Van Hove, *Physica*, **21** (1955), 517.

previously discussed mechanical features. Ohm's law states in addition that the drift velocity acquired is proportional to the applied field. This means that the frictional force is highly reproducible in a given sample, and can perhaps be understood by treating the field as a first order perturbation on the equilibrium state of the electric carriers.

CLASSICAL RELAXATION THEORY OF COLLISIONS; MOBILITY

The picture of conduction electrons which evolved in the last two chapters is essentially one of a gas. The particles of the gas obey an unusual kind of dynamics, which is determined by the structure of the band to which they belong. The adaptation of kinetic theory to this new kind of dynamics is easy. A more awkward question is what the nature of the collisions are, and whether the usual postulates of kinetic theory are applicable to them. The answer to this question is difficult even for gases; the kinetic formalism itself shows that it is not applicable to Coulomb forces because of their long range; however, even there, semiquantitative results may be obtained by patching up the theory. It is reasonable, therefore, to disregard this question at the present time and to apply the apparatus of the kinetic theory of gases to particles having a hamiltonian of the form (6.53). The basic element of this apparatus is a distribution function which gives the probability of finding a given position and velocity realized among the particles; the variation of this function in time is governed by the *Boltzmann transport equation*. In the present instance, we deal with motion of electrons confined to certain bands; the conjugate variables $\hbar\mathbf{k}$ and $\mathbf{r}$ are therefore the correct ones to use together with the band number q, assumed fixed. We use these variables in the quasi-classical sense evolved in the last chapter. Our distribution function f has thus the form

$$f = f(\mathbf{k}; \mathbf{r}) \tag{7.1}$$

and the transport equation for it reads†

$$\left(\frac{df}{dt}\right)_{\text{path}} = \left(\frac{\partial f}{\partial t}\right)_{\text{coll}}. \tag{7.2}$$

† See S. Chapman and T. G. Cowling, *The Mathematical Theory of Non-uniform Gases* (Cambridge, University Press, 1952), ch. 3.

Here the left side is the total time derivative along a hypothetical trajectory in phase space, which is calculated neglecting collisions, and the right-hand side is the change of f due to collisions. The major difficulty of kinetic theory is the complicated structure of this term. In the present instance we may perhaps neglect the transitions from band to band. We may also presume that the term is linear in f because electron-electron collisions preserve momentum and thus are ruled out in a theory of Ohm's law. However, these features are not enough to simplify the problem, and in fact a very careful study of equation (7.2) would show that it cannot be rigorously implemented. The equation is itself already an approximation, and we must expect to make approximations in evaluating its right-hand term. Having this firmly in mind, we can get a clue to the form of this term by studying the models for which (7.2) is rigorously true: these models assume collisions to be instantaneous events. The term decomposes then into two parts: a term $-f/\tau$ which describes the removal of particles from a portion of phase space due to collisions, and another term which describes their reappearance at other positions. The factor τ in the first term is the mean free time between collisions; it will in general be a function of q and $\mathbf{k}$, but rarely of $\mathbf{r}$; the second term is actually a complicated sum of integrals which is homogeneous of the first degree in the f's and cancels with $-f/\tau$ at thermal equilibrium. The tendency of the two terms taken together is therefore to make any distribution evolve toward thermal equilibrium at a rate which is of the order $1/\tau$. This effect is reproduced by the following rendition of the term

$$\left(\frac{\partial f}{\partial t}\right)_{\text{coll}} = -\frac{f(\mathbf{k};\mathbf{r})-f_0(\mathbf{k};\mathbf{r})}{\tau(\mathbf{k})}. \qquad (7.3)$$

Here f_0 is a function which is spatially identical with f and yields an average energy which is shifted only slightly, but equals the Fermi–Dirac function (5.62) in its dependence on $\mathbf{k}$; this Fermi–Dirac function can be replaced by its Boltzmann approximations (5.64) or (5.65) if the carriers are non-degenerate. The *relaxation time* $\tau(\mathbf{k})$ entering into (7.3) may have to be taken slightly longer than the time between collisions introduced previously, to take care of 'memory' during collisions. Experience has shown that the error committed in adopting the form (7.3) is usually small. With (7.3), equation (7.2) takes the form

$$\frac{\partial f}{\partial t}+\frac{d\mathbf{k}}{dt}\cdot\frac{\partial f}{\partial \mathbf{k}}+\frac{d\mathbf{r}}{dt}\cdot\frac{\partial f}{\partial \mathbf{r}}+\frac{f-f_0}{\tau(\mathbf{k})}=0,$$

or with (6.54) and (6.55)

$$\frac{\partial f}{\partial t}-\frac{1}{h}\frac{\partial V}{\partial \mathbf{r}}\cdot\frac{\partial f}{\partial \mathbf{k}}+\frac{1}{h}\frac{\partial W}{\partial \mathbf{k}}\cdot\frac{\partial f}{\partial \mathbf{r}}+\frac{f-f_0}{\tau}=0. \tag{7.4}$$

Let us apply (7.4) first to the basic case of Ohm's law, that is the case in which $\mathscr{H}$ has the form (6.62). We get then

$$\frac{\partial f}{\partial t}+\frac{e\mathbf{E}}{h}\cdot\frac{\partial f}{\partial \mathbf{k}}+\frac{1}{h}\frac{\partial W}{\partial \mathbf{k}}\cdot\frac{\partial f}{\partial \mathbf{r}}+\frac{f-f_0}{\tau}=0. \tag{7.5}$$

In the present instance, f is a constant in space and time. In its dependence on $\mathbf{E}$ it can be considered a power series, with f_0 the constant term. The linear term is then responsible for Ohm's law, which appears thus as a limiting law for a small electric field $\mathbf{E}$. In practice, this restriction has little importance.† For our purposes we may, therefore, solve (7.5) approximately, dropping the first and the third term, and replacing f by f_0 in the second. This yields for f

$$f(\mathbf{k})=f_0(\mathbf{k})-\frac{eE\tau}{h}\frac{\partial f_0}{\partial k_x}, \tag{7.6}$$

where $\mathbf{E}$ is assumed to point along the x-direction. We finally compute the mean drift velocity $\langle\mathbf{v}\rangle$, using (6.24) and introducing the carrier density n:

$$n\langle\mathbf{v}\rangle=\frac{1}{h}\int f(\mathbf{k})\frac{\partial W}{\partial \mathbf{k}}\,d\mathbf{k}.$$

The first term of (7.6) makes no contribution to this average; we get therefore

$$n\langle\mathbf{v}\rangle=-\frac{eE}{h^2}\int\tau(\mathbf{k})\frac{\partial f_0}{\partial k_x}\frac{\partial W}{\partial \mathbf{k}}\,d\mathbf{k}. \tag{7.7}$$

An integration by parts brings (7.7) into the form

$$n\langle\mathbf{v}\rangle=\frac{eE}{h^2}\int\frac{\partial}{\partial k_x}\left(\tau(\mathbf{k})\frac{\partial W}{\partial \mathbf{k}}\right)f_0(\mathbf{k})\,d\mathbf{k}.$$

† In semiconductors, the high built-in electric fields of n-p junctions permit observation of the higher terms under favourable conditions. See W. Shockley *Bell System Tech. J.* **30** (1951), 990.

The expression on the right is now an average over the unperturbed thermal distribution f_0; hence it can be written as

$$\langle \mathbf{v} \rangle = \frac{eE}{\hbar^2} \left\langle \frac{\partial}{\partial k_x} \left(\tau(\mathbf{k}) \frac{\partial W}{\partial \mathbf{k}} \right) \right\rangle, \qquad (7.8)$$

where the pointed bracket indicates a thermal average over the unperturbed distribution. The simplest sub-case of (7.8) arises when the relaxation time τ is a constant and $W(\mathbf{k})$ has the form (5.61). We get then

$$\langle v_x \rangle = \frac{eE\tau}{m^*}. \qquad (7.9)$$

In a loose way of speaking we can say in all cases that the result for the drift velocity is (7.9), since it is dimensionally identical with (7.8). In this sense (7.8) simply specifies which particular type of average of the quantities in (7.9) actually enters into (7.9). In addition (7.9) has obvious advantages for qualitative reasoning.

Accepting (7.9) as the result in the sense stated, we see that it yields a drift velocity proportional to the electric field. This is the atomic form of Ohm's law. The constant of proportionality is called the *mobility* of the carriers. Equation (7.9) yields for the electron mobility μ the expression

$$\mu = \frac{e\tau}{m^*}. \qquad (7.10)$$

The transition from a mobility to a conductivity is made via equation (6.69) and the definition (6.68). We find for the conductivity σ

$$\sigma = \frac{ne^2\tau}{m^*}. \qquad (7.11)$$

The expressions (7.7)–(7.11) provide the necessary connecting link between a hypothetical relaxation process to which the electrons are subject and their ohmic behaviour under an applied electric field. Naturally, for a complete calculation we must compute τ from first principles; this will be done below for a few selected cases. In the meantime, certain general features are apparent from the relations as they stand. First, we see from (7.8) that the mobility is a material tensor characteristic of the crystal. Second, we see that the tensor is symmetric; for the expression (7.7) is symmetric, because f_0 depends on $\mathbf{k}$ only through the intermediary of $W(\mathbf{k})$. Third, the contribution to the charge transport arises entirely from regions in **k**-space in which f varies strongly with W; therefore, if we deal with a degenerate Fermi gas of electrons then the motion of charges under the influence of an electric field is limited to those near the Fermi level. In fact the bulk of the quasi-free electrons of a metal are inert to all disturbances involving small energy changes, as was pointed out in Chapter 5. What is curious is that, in the case of thermal agitation, this effect produces a reduction of the specific heat, as exhibited in equation (5.79), while for the bulk conductivity this is not the case; indeed the formulas (7.8)–(7.11) show no essential trace of the statistics to which the electrons are subject.

A second transport phenomenon of importance is the diffusion of electrons in solids. Rather than treating this as an independent phenomenon let us establish the close relationship between diffusion and electrical conduction, which exists for non-degenerate electrical carriers through the so-called *Einstein relation*†. To do this, we assume in equation (7.5) the simultaneous presence of a constant electric field **E** and a constant relative concentration gradient α along the x direction. In other words, f is now taken of the form

$$f(\mathbf{r}; \mathbf{k}) = e^{\alpha r_x} G(\mathbf{k}). \tag{7.12}$$

As was discussed in connexion with equation (7.3), $f_0(\mathbf{r}; \mathbf{k})$ has the same spatial dependence. Its dependence on **k**, on the other hand, is governed by equilibrium considerations. The equilibrium distribution is the Fermi–Dirac distribution in one of its simplified forms (5.64) or (5.65). We can write therefore

$$f_0(\mathbf{r}; \mathbf{k}) = e^{\alpha r_x} e^{-W(\mathbf{k})/kT}. \tag{7.13}$$

All terms are now non-vanishing in (7.5) except the first. If only linear terms in α and E are retained then we can solve for f as previously, and we get, with (7.12) and (7.13),

$$f = f_0 + \frac{1}{h}\frac{\partial W}{\partial k_x}\left(\alpha - \frac{e\mathrm{E}}{kT}\right)f_0$$

or

$$f = f_0 + \frac{1}{h}\frac{\partial W}{\partial \mathbf{k}} \cdot \left(\alpha - \frac{e\mathbf{E}}{kT}\right)f_0. \tag{7.14}$$

---

† The relation was first derived by W. Nernst, *Zeits. f. Physik. Chem.* 9 (1884), 613.

In other words, a concentration gradient and a field are equivalent since they only enter through the large bracket in (7.14). If we now denote by ρ the particle density and use the definition of the diffusion tensor (D) then we have in the presence of a density gradient alone from (7.13)

$$\mathbf{i} = -(D)\frac{\partial \rho}{\partial \mathbf{r}} = -(D)\alpha\rho.$$

In the presence of a field we substitute for α the large bracket of (7.14) and get

$$\mathbf{i} = -(D)\alpha\rho + \frac{e(D)}{kT}\mathbf{E}\rho. \tag{7.15}$$

By definition the multiplier of $\mathbf{E}\rho$ in the expression for the current density is the mobility tensor (μ). Because of (7.15), it is related to the diffusion tensor through

$$(\mu) = \frac{e}{kT}(D). \tag{7.16}$$

This is the relation which was to be established.

The previous calculation of Ohm's law can be extended to the case of an alternating electric field. We take $\mathbf{E}$ of the form

$$E_x = E_0 e^{i\omega t} \tag{7.17}$$

and return again to (7.5), this time asking for an f varying in time but spatially constant. If we look for an f varying with the same frequency as E, we find previous results maintained except that we must everywhere use the substitution

$$\frac{1}{\tau} \rightarrow \frac{1}{\tau} + i\omega. \tag{7.18}$$

We shall write down the alternating current conductivity only for the special case that the simplifications underlying (7.11) apply. We get then

$$\sigma = \frac{ne^2}{m^*}\frac{(1/\tau) - i\omega}{(1/\tau^2) + \omega^2}. \tag{7.19}$$

Thus, conductivity decreases with increasing frequency and finally approaches a condition where τ disappears from the equation. The limiting formula arises from the oscillations of quasi-free electrons in a field and could have been derived by the methods of the previous chapter. For this reason, the simplicity of the result persists even in the general situation exemplified by (7.8). Let us define the reciprocal effective mass tensor by the logical extension of (6.50), namely

$$\left(\frac{1}{m}\right) = \left(\frac{1}{m}\right)_{ij} = \frac{1}{\hbar^2}\left\langle\frac{\partial^2 W}{\partial k_i\,\partial k_j}\right\rangle, \tag{7.20a}$$

where the average is taken over the equilibrium thermal distribution of the carriers; then with (7.18), equation (7.8) takes the limiting form

$$\langle\mathbf{v}\rangle = -\,\mathrm{i}\frac{e}{\omega}\left(\frac{1}{m}\right)\mathbf{E}; \tag{7.20b}$$

this is just what would result from the purely dynamical considerations of the last chapter. We are touching here closely the theory of the electric susceptibility of solids; indeed, the result (7.19) can also be looked at as a formula for the dielectric constant, in which case real and imaginary parts exchange their roles.

There is a practical difficulty about observing (7.19) for electrons in solids, namely the modifications of the direct current resistance by effects of electromagnetic origin, such as inductance, skin effect, etc. These effects occur at much lower frequency than the one discussed here because τ for electrons is in the neighbourhood of 10^{-12} sec. Thus the formula derived here has to be checked by optical measurements in the infra-red rather than by electrical methods.

QUANTUM RELAXATION THEORY OF COLLISIONS; LINE BROADENING

Formula (7.19) is typical for the response of a relaxing system to a simple harmonic stimulation. The next item in order of complication is the study of a resonant response. This could be done within a classical framework by taking up again a phenomenon such as diamagnetic resonance studied in the previous chapter. However, it would be very unsatisfactory to treat this classically, because classical cyclic motions are rarely free of overtones, while according to Bohr's correspondence principle each quantum transition is equivalent to a classical harmonic oscillator having no overtones. Hence there is a fundamental simplicity to the quantum problem of the influence of relaxation processes upon transitions between energy states, which is absent from all but the simplest classical models.

The natural quantum generalization of the distribution function of Boltzmann is a table of numbers P_n giving the probability of finding an individual carrier in the quantum state n. Since P_n is the absolute square of an amplitude with phase it does not carry the full information necessary to determine the behaviour of the system of carriers for all time. We can, however, embed the numbers P_n as diagonal elements into a matrix whose general element is

$$\rho_{mn} = A_m A_n^*. \tag{7.21}$$

where A_m is the amplitude of the participating state m. This is the most elementary definition one can give of the *density matrix* ρ. From it, one can prove by elementary quantum mechanical methods that it has the following properties:

(a) It is hermitian.

(b) It transforms like a dynamical variable under a change of the base.

(c) It obeys the equation of motion of a dynamical variable:

$$i\hbar \frac{d\rho}{dt} = \mathcal{H}\rho - \rho\mathcal{H}. \tag{7.22}$$

It is interesting that equation (7.22) allows for the matrix ρ a choice without the off-diagonal elements implicit in the original definition (7.21), namely

$$\rho_{nm} = P_n \delta_{n,m}, \tag{7.23}$$

where the indices n and m refer to the eigenstates of $\mathcal{H}$. The way to understand this is to remember that the off-diagonal members of (7.21) have rapidly varying time factors of the form

$$\text{ixp} \frac{1}{\hbar}(E_m - E_n)(t - t_0)$$

and that if we have a statistical ensemble of systems (in our case a number of almost identical crystals) each one may have a different phase constant t_0. If the value of this phase is random, then the off-diagonal elements vanish for the ensemble and the diagonal elements become the appropriate probabilities in statistical equilibrium. This is the physical meaning of (7.23).

In order to investigate the influence of collisions on the quantum jumps of excess of defect electrons we must presume to have a hamiltonian having three distinct well-defined parts:

$$\mathcal{H} = \mathcal{H}^{\text{I}} + \mathcal{H}^{\text{II}} + \mathcal{H}^{\text{III}}. \tag{7.24}$$

I. $\mathcal{H}^{\mathrm{I}}$ is a large time-independent part which specifies the eigenstates of the carriers.

II. $\mathcal{H}^{\mathrm{II}}$ is a small 'externally-imposed' part, varying at a fixed frequency ω which induces transitions between the levels I.

III. $\mathcal{H}^{\mathrm{III}}$ is the collision part which makes any distribution of the carriers among the levels I relax toward the thermal equilibrium (7.23).

The first term $\mathcal{H}^{\mathrm{I}}$ will be used to determine, through its eigenstates, the representation employed for ρ. We shall write

$$\langle n\,|\,\mathcal{H}^{\mathrm{I}}\,|\,m\rangle = \delta_{n,m}E_n \tag{7.25a}$$

$$\hbar\omega_{nm} = E_n - E_m. \tag{7.25b}$$

By itself $\mathcal{H}^{\mathrm{I}}$ admits for ρ any solution either of the form (7.21) or (7.23). The term $\mathcal{H}^{\mathrm{II}}$ inducing transitions will be written as

$$\mathcal{H}^{\mathrm{II}} = V\cos\omega t, \tag{7.26}$$

where V is independent of time. We may also assume the matrix elements of V real, which is always possible if V is real and the degenerate wave functions of $\mathcal{H}^{\mathrm{I}}$ are chosen accordingly. The collision term $\mathcal{H}^{\mathrm{III}}$ is the term causing the greatest difficulty, as in the previous classical treatment by the Boltzmann equation. In the first place, there is no clear principle which allows us to distinguish $\mathcal{H}^{\mathrm{III}}$ from $\mathcal{H}^{\mathrm{I}}$. Furthermore, its structure must be very complicated since it is meant to represent random collisions. The expedient analogous to (7.3) is therefore almost inevitable: we must try to approximate $\mathcal{H}^{\mathrm{III}}$ from its effects. We know in this respect that its net result must be linear in ρ, and of such a nature that it makes any ρ relax toward the form (7.23), with the quantities P_n having their thermal equilibrium values. Whether the equilibrium should be the instantaneous equilibrium in $\mathcal{H}^{\mathrm{I}}$ and $\mathcal{H}^{\mathrm{II}}$ or just the equilibrium in $\mathcal{H}^{\mathrm{I}}$ is another difficult question; which answer is the right one depends on the rapidity of the relaxation rate as compared to the frequency ω in (7.26). We shall presume here the system to relax toward (7.25). The simplest form consistent with these ideas is the analogue of (7.3) proposed by Karplus and Schwinger namely[†]

$$\left(\frac{\mathrm{d}\rho}{\mathrm{d}t}\right)_{\mathcal{H}^{\mathrm{III}}} = -\frac{1}{\tau}(\rho - \rho^E), \tag{7.27}$$

[†] R. Karplus and J. Schwinger, *Phys. Rev.* **73** (1948), 1020.

where ρ^E equals the equilibrium value (7.23) of ρ. (7.27) thus replaces for the collision part $\mathscr{H}^{\mathrm{III}}$ the formulation (7.22) of which it represents an approximate time average over several impacts.

As a result of these several substitutions equation (7.22) takes the form

$$\frac{d\rho_{nm}}{dt} = -i\omega_{nm}\rho_{nm} - \frac{i}{\hbar}\cos\omega t \sum_{\nu}(V_{n\nu}\rho_{\nu m} - \rho_{n\nu}V_{\nu m}) - \frac{1}{\tau}(\rho_{nm} - \rho_{nm}^E).$$

(7.28)

The equation may be solved with $\mathscr{H}^{\mathrm{II}}$ treated as a perturbation, as in the corresponding classical situation (7.6). ρ_{nm} differs then very little from ρ_{nm}^E; this means also that the diagonal elements of ρ are large compared to the others. If this is true and if the matrix elements V_{nm} do not show a compensating irreguliarity in size (such as selection rules), then within the perturbation term of (7.28) the two terms involving diagonal elements of ρ are large, and the others may be neglected; also those elements may be replaced by their equilibrium values P_n^E. This yields for the off-diagonal elements, $n \neq m$:

$$\frac{d\rho_{nm}}{dt} + i\omega_{nm}\rho_{nm} + \frac{1}{\tau}\rho_{nm} = -\frac{i}{\hbar}V_{nm}(P_m^E - P_n^E)\cos\omega t. \quad (7.29)$$

This is a straightforward equation for a harmonically driven linear system and yields

$$\rho_{nm} = \frac{V_{nm}}{2\hbar}(P_n^E - P_m^E)\left\{\frac{e^{i\omega t}}{\omega + \omega_{nm} - (i/\tau)} - \frac{e^{-i\omega t}}{\omega - \omega_{nm} + (i/\tau)}\right\}. \quad (7.30)$$

For the diagonal elements $\rho_{nn} = P_n$, the calculation runs slightly differently because the 'large' term in the perturbation sum of (7.28) is missing. We must therefore keep all terms of the equation (7.28), but can substitute (7.30) for the off-diagonal elements in it. This yields

$$\frac{dP_n}{dt} + \frac{1}{\tau}(P_n - P_n^E) = -\frac{i}{\hbar}\sum_{\nu}V_{n\nu}(\rho_{\nu n} - \rho_{n\nu})\cos\omega t =$$

$$= -\frac{i}{2\hbar^2}\sum_{\nu}V_{\nu n}^2(P_\nu^E - P_n^E)\cos\omega t \times$$

$$\times\left\{\frac{e^{i\omega t}}{\omega + \omega_{\nu n} - (i/\tau)} + \frac{e^{i\omega t}}{\omega - \omega_{\nu n} - (i/\tau)} - \frac{e^{-i\omega t}}{\omega + \omega_{\nu n} + (i/\tau)} - \frac{e^{-i\omega t}}{\omega - \omega_{\nu n} + (i/\tau)}\right\}.$$

It is seen that P_n oscillates at the frequency 2ω about a stationary value $\langle P_n \rangle$ which is different from P_n^E. One finds its value by collecting constant terms on the right; it equals

$$\langle P_n \rangle - P_n^E = \frac{1}{2\hbar^2} \sum_\nu V_{\nu n}^2 (P_\nu^E - P_n^E) \times$$

$$\times \left\{ \frac{1}{(\omega - \omega_{\nu n})^2 + (1/\tau^2)} + \frac{1}{(\omega + \omega_{\nu n})^2 + (1/\tau^2)} \right\}. \quad (7.31)$$

From this result we can reason further on the transition rate thus: since, according to (7.27), the relaxation rate proceeds as

$$\frac{1}{\tau}(\langle P_n \rangle - P_n^E),$$

the transition rate due to V must also have that value and proceed in the inverse direction so as to maintain the stationary state.

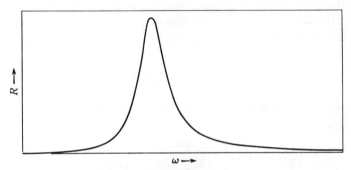

Fig. 7.1. Shape of a collision-broadened line (Lorentz line shape).

Finally, since the power absorbed per transition equals $\hbar\omega$, the rate R of power absorption must equal $\hbar\omega/\tau$ times the expression (7.31):

$$R = \frac{\omega}{2\hbar\tau} \sum_\nu V_{\nu n}^2 (P_\nu^E - P_n^E) \left\{ \frac{1}{(\omega - \omega_{\nu n})^2 + (1/\tau^2)} + \frac{1}{(\omega + \omega_{\nu n})^2 + (1/\tau^2)} \right\}. \quad (7.32)$$

If we study the spectral response line by line, that is, consider at one time only transitions caused by one spectral line ω_{mn}, then we see that the relaxation mechanism has *broadened* the line to a characteristic shape with wide wings. The shape of the spectral response curve (7.32) is shown in fig. 7.1. It is characteristic of collision-broadened lines. The broadening does not affect the total transition rate R' attributable to the matrix element V_{mn}; indeed we find

$$R' = \int_0^\infty \frac{1}{\tau}\{\langle P_n\rangle - P_n^E\}\,d\omega = \frac{1}{2\hbar^2\tau}\,V_{mn}^2(P_m^E - P_n^E) \times$$

$$\times \int_0^\infty d\omega \left\{\frac{1}{(\omega-\omega_{mn})^2+(1/\tau^2)} + \frac{1}{(\omega+\omega_{mn})^2+(1/\tau^2)}\right\},$$

or

$$R' = \frac{\pi}{2\hbar^2}\,V_{mn}^2(P_m^E - P_n^E). \tag{7.33}$$

Equation (7.33) is a particular instance of the *principle of spectroscopic stability*, which states that the total transition rate between two terms is, for given population, not affected by perturbations splitting or broadening the terms. The principle is a rule rather than a law of nature: if one of the perturbations involved gets sufficiently powerful to mix terms with each other the principle ceases to be valid.

It is not surprising that the simple rendition (7.27) of a quantum relaxation process is not adequate for many situations. It would be consistent with the assumed formalism to take for $1/\tau$ a matrix multiplying $(\rho-\rho^E)$. A more limited generalization, which has great importance, consists in making a distinction between the way diagonal and off-diagonal elements relax. It was pointed out in connexion with equation (7.23) that the latter have phase factors varying rapidly in time, while the diagonal elements have no such factors. These factors might be expected to contribute to the relaxation process if the system loses phase memory in time, and it is therefore not surprising that off-diagonal elements have usually a larger relaxation rate than diagonal ones. It is this relaxation rate which controls the line shape. Consequently, the relaxation rate determined from line shape usually comes out larger than the one observed in an actual relaxation experiment.

SCATTERING PROCESSES FOR ELECTRONS; MOBILITY FORMULA FOR THERMAL SCATTERING

Having formally analysed the effect of a relaxation mechanism on the response of quasi-free electrons in solids, we are left with the task of identifying these processes in actual materials. Generally speaking, the processes must represent some sort of scattering. The most obvious idea for explaining the origin of the scattering is to blame the atoms which make up the solid. These atoms, one might think, act as scattering centres for the electrons travelling through the lattice. The previous two chapters have refuted this notion: a perfectly regular array of atoms does not constitute an obstacle to the flow of electrons. A second form of scattering which we must rule out are the collisions of electrons with each other. Such collisions play a predominant role in the kinetic theory of gases. In the case of Ohm's law these collisions can be ruled out on a priori grounds, because they do not check the motion of the gas as a whole; this type of check is needed for implementing equation (7.3). We have therefore to fall back on features of the solid matrix which may produce scattering because they were ignored in the electron dynamics of the preceding chapter. These features are irregularities of any kind: impurities, voids, lattice misfits and distortions. It is useful to divide them into two categories: frozen-in structural crystal imperfections and rapidly varying distortions arising from thermal agitation; this second kind will be present even in a perfect crystal.

The frozen-in imperfections are the conceptually simpler thing to discuss. If we deal with centres which are well separated from each other then the concept of a cross-section ought to apply. We say then that the centres present a scattering cross-section σ to the electric carriers from which τ is derived by

$$\tau = \frac{1}{\rho \sigma \sqrt{\langle v^2 \rangle}}. \tag{7.34}$$

Here ρ denotes the density of scattering centres, and $\sqrt{\langle v^2 \rangle}$ is the mean random velocity of the carriers. We may even be lucky enough to have scattering centres whose cross-section is sensibly independent of the carrier velocity. This, in turn, yields through (7.10) and (7.34) a mobility varying as $1/\sqrt{\langle v^2 \rangle}$. In metals this makes it a constant, because the random velocity of electrons in metals is of the 'zero point' variety and hence independent·of temperature; this is indeed the experimental result for metals at low temperature. For semiconductors, on the other hand, this variation as $1/\sqrt{\langle v^2 \rangle}$ would yield a variation as $1/T$. This may happen occasionally, but is not the most common result. The most common case is embodied in the Conwell–Weisskopf formula which describes scattering due to point charges.† The analysis proceeds from Rutherford's law of scattering.

Scattering by frozen-in imperfections is generally a low-temperature effect. At moderate and high temperature scattering by the thermal agitation of the lattice is predominant. The cross-section concept of (7.34) is not quite applicable here because lattice waves are everywhere. The derivation of the mean free path for electrons produced by thermal agitation of the lattice will be given now. For convenience, it will be divided into three parts:

(1) Calculation of the matrix elements of the coupling energy between the modes of the electronic motion which arise from lattice waves of given amplitude and wave length.

(2) Analysis of the relaxation process for the carriers produced by an interaction of the type (1).

(3) Evaluation of the vibrational amplitudes of the lattice in thermal equilibrium.

(1) The coupling between electrons and lattice waves really proceeds through absorption and emission of phonons. However, for the discussion of induced absorption and emission, the field properties of the phonon field are not essential (as is the case for the analogous case in optics). Furthermore, the frequencies of the sound waves are not essential for the present purpose, and will be neglected below, because they are much smaller than the frequencies of the electron waves, and thus lead only to a small Raman effect, that is a small change in the energy of the electrons; it is their angular scattering in $\mathbf{k}$-space, which is the fastest process and controls the relaxation of a non-equilibrium electronic distribution. Furthermore, we may, in our sample calculation, assume for the coupling the simple formula (3.73) which results from the deformation potential theory. With all these assumptions, the interaction energy is a sum of terms of the form (3.74), namely

$$U = 2\pi u \sum_{\varkappa, \nu} a_\nu(\varkappa)\, \mathbf{e}_\nu \cdot \varkappa \sin 2\pi \varkappa \cdot \mathbf{r}, \qquad (7.35)$$

† E. Conwell and V. F. Weisskopf, *Phys. Rev.* **77** (1950), 388. S. Chapman, *Monthly Notices R.A.S.* **82** (1922), 292.

where ν runs over all modes existing for fixed κ: polarizations, as exhibited in (3.36), optical and acoustical branches, etc. In formula (7.35), the lattice vector $\mathbf{r}$ has been used rather than the co-ordinate $\mathbf{x}$; this is not entirely arbitrary. First of all, $\mathbf{r}$ may be more closely right than $\mathbf{x}$, if a 'rigid cell' rather than a 'deformable cell' approaches more closely to the truth. Second, even if $\mathbf{x}$ were right it would only differ essentially from $\mathbf{r}$ in its interband effects which are neglected throughout in this chapter.† Thus (7.35) may be taken as the interaction, with $\mathbf{r}$ having the technical meaning given to it in Chapter 6. For the transformation of the matrix element of U to a Bloch base we use equation (6.12); the summations over $\mathbf{r}$ and $\mathbf{r}'$ occurring in that equation can be done explicitly for the expression (7.35) and yield

$$\langle \mathbf{k} \,|\, U \,|\, \mathbf{k} - \kappa \rangle = -i\pi u \sum_{\nu} \{a_{\nu}(\kappa) + a_{\nu}(-\kappa)\} \mathbf{e}_{\nu} \cdot \kappa, \qquad (7.36)$$

where the sum over ν runs again over the small number of terms enumerated in connection with (7.35).

(2) We shall now derive, rather than assume, the relaxation time τ assumed in (7.27). In doing this we shall again employ the density matrix formalism (7.22), but shall, in accordance with the remarks made at the beginning of the chapter, give to $\mathscr{H}^{\mathrm{III}}$ a different status from $\mathscr{H}^{\mathrm{I}}$: $\mathscr{H}^{\mathrm{I}}$ determines the eigenstates, and $\mathscr{H}^{\mathrm{III}}$ is assumed to cause transitions between them. As initial condition we assume that

$$\rho_{nm} = \delta_{n,0} \delta_{m,0}, \qquad (7.37a)$$

where the index 0 corresponds to a particular wave vector $\mathbf{k}$:

$$\mathbf{k} = \mathbf{k}_0, \qquad (7.37b)$$

which is realized initially with unit probability. The hamiltonian has the form

$$\mathscr{H} = \mathscr{H}^{\mathrm{I}} + U, \qquad (7.38)$$

that is, the term $\mathscr{H}^{\mathrm{II}}$ of (7.24) is absent, and $\mathscr{H}^{\mathrm{III}}$ is specified precisely by its matrix elements (7.36). The form (7.27) is expected

† One can show explicitly that if the crystal has a centre of symmetry the multipliers of the Bloch functions $b_q(\mathbf{x}; \mathbf{k})$ can be chosen in such a way that all matrix elements $\langle q; \mathbf{r} \,|\, \mathbf{x} - \mathbf{r} \,|\, q; \mathbf{r}' \rangle$ are zero.

as the result of a calculation which employs only (7.22). Thus, with (7.25), we start with

$$\frac{d\rho_{nm}}{dt} = -i\omega_{nm}\rho_{nm} - \frac{i}{\hbar}\sum_{\nu}(U_{n\nu}\rho_{\nu m} - \rho_{n\nu}U_{\nu m}). \qquad (7.39)$$

Since all the initial strength is in ρ_{00} and its coupling is exclusively to elements ρ_{0n} and ρ_{n0} we may neglect the other elements in estimating the early linear drop of ρ_{00} in time. Also we shall follow the drainage process for a time which is sufficiently short so that ρ_{00} remains substantially equal to 1 and the other elements remain small. Under these conditions, the general equation system (7.39) takes the simplified form

$$\frac{d\rho_{0m}}{dt} = i\omega_{m0}\rho_{0m} + \frac{i}{\hbar}U_{0m}, \qquad (7.40a)$$

$$\frac{d\rho_{n0}}{dt} = -i\omega_{n0}\rho_{n0} - \frac{i}{\hbar}U_{n0}, \qquad (7.40b)$$

$$\frac{d\rho_{00}}{dt} = -\frac{i}{\hbar}\sum_{\nu}(U_{0\nu}\rho_{\nu 0} - \rho_{0\nu}U_{\nu 0}). \qquad (7.40c)$$

This system can be solved directly by solving first the equations (a) and (b) with the initial condition (7.37) and inserting into (c). Integration of this last equation is not needed since $d\rho_{00}/dt$ is just what we wish to find. It comes out to be

$$\frac{d\rho_{00}}{dt} = -\frac{2}{\hbar^2}\sum_{\nu}\frac{|U_{\nu 0}|^2}{\omega_{\nu 0}}\sin\omega_{\nu 0}t. \qquad (7.41)$$

The most important feature of (7.41) is the negative sign. The constancy of the drainage from ρ_{00} does not show up until the summation over ν is carried out and t is permitted to grow sufficiently big to shrink the integral over the final states to those which conserve energy. The integral is then a Dirichlet integral of the type

$$\lim_{t=\infty}\int_{\omega_1}^{\omega_2}f(\omega_{\nu 0})\frac{\sin\omega_{\nu 0}t}{\omega_{\nu 0}}d\omega_{\nu 0} = \pi f(0) \quad \text{if} \quad \omega_1 < 0 < \omega_2.$$

In terms of the δ-function symbolism this means

$$\frac{\sin\omega_{\nu 0}t}{\omega_{\nu 0}} \approx \pi\delta(\omega_{\nu 0}),$$

and hence from (7.41)

$$\frac{d\rho_{00}}{dt} = -\frac{2\pi}{\hbar^2} \sum_\nu |U_{\nu0}|^2 \,\delta\omega_{\nu0}. \tag{7.42}$$

The formula may be found in standard text-books. It is held to prove the exponential decay rate of ρ_{00}, because it holds only as long as ρ_{00} remains close to 1, and in that case we have approximately,

$$1 - \frac{t}{\tau} \approx e^{-t/\tau}.$$

Hence we get from (7.42)

$$\frac{1}{\tau} = \frac{2\pi}{\hbar^2} \sum_\nu |U_{\nu0}|^2 \,\delta\omega_{\nu0}. \tag{7.43}$$

The sum is one of those subject to the identity (3.41). Inserting it into (7.43) we get for $1/\tau$ the final formula

$$\frac{1}{\tau(\mathbf{k}_0)} = \frac{2\pi\Omega}{\hbar} \iiint |\langle \mathbf{k}_0 | U | \mathbf{k}\rangle|^2 \, d\mathbf{k} \,\delta\{W(\mathbf{k}) - W(\mathbf{k}_0)\}. \tag{7.44}$$

(3) An easy way to carry out the third part of the derivation is to compute the mean kinetic energy in a wave of the type (3.66) and setting it, because of equipartition, equal to half the thermal energy (3.56). For simplicity we shall also assume that the acoustical branch is predominant in the scattering. This is often the case at sufficiently high temperature, so that kT is large compared to an acoustical phonon energy but not yet large compared to a quantum of the upper branches. (3.56) reduces then to $\langle E \rangle = kT$. With these simplifications we get from (3.66) for the kinetic energy K

$$K = \tfrac{1}{2}m_c \sum_\mathbf{r} (\dot{\boldsymbol{\xi}}(\mathbf{r}))^2 = 2\pi^2 m_c f^2 a^2 \sum_\mathbf{r} \sin^2 2\pi(\boldsymbol{\kappa}\cdot\mathbf{r} - ft)$$

$$= \pi^2 m_c f^2 a^2 N = \tfrac{1}{2}kT,$$

where m_c is the mass of one primitive cell. This yields for a^2

$$a^2 = \frac{kT}{2\pi^2 f^2 \rho\Omega}, \tag{7.45}$$

where ρ is the mass density of the crystal. It is seen that the amplitude is primarily a function of the frequency f rather than the wave vector $\mathbf{k}$. In practice it is often possible to distinguish longitudinal and transverse waves in a crystal and to discard the latter because of the factor $\mathbf{e}_\nu \cdot \mathbf{\kappa}$ in (7.36). For the former, an approximate velocity of propagation c_L can be introduced, yielding

$$(a(\mathbf{\kappa}))^2 = \frac{kT}{2\pi^2 \rho \Omega \kappa^2 c_L^2}. \tag{7.46}$$

The results of the three parts of the reasoning, equations (7.36), (7.44) and (7.46) may now be combined to yield an expression for the relaxation time of electrons which arises from thermal scattering. In squaring (7.36) and inserting (7.46) into it, one is to remember that the mean product of two amplitudes belonging to different normal modes is always zero. One gets therefore

$$|\langle \mathbf{k}_0 | U | \mathbf{k}_0 - \mathbf{\kappa} \rangle|^2 = \frac{u^2 kT}{\rho \Omega c_L^2}$$

which is the form required for insertion into (7.44). It is seen that the matrix element comes out independent of $\mathbf{k}_0$ and $\mathbf{\kappa}$, so that it can be taken out of the integration sign. (7.44) becomes therefore

$$\frac{1}{\tau(\mathbf{k}_0)} = \frac{2\pi u^2 kT}{\rho \hbar c_L^2} \iiint d\mathbf{k}' \delta\{W(\mathbf{k}') - W(\mathbf{k}_0)\}. \tag{7.47}$$

(7.47) is the final result for a general band shape. As usual, its interpretation is facilitated by taking the case (5.61) of a simple effective mass relation. The integral is then elementary, and we get

$$\frac{1}{\tau(\mathbf{k}_0)} = \frac{2m^* u^2 kT k_0}{\hbar^3 \rho c_L^2}. \tag{7.48}$$

It is seen that τ is inversely proportional to k_0 rather than a constant as assumed previously. A mean free path λ independent of k_0 can therefore be defined through (5.59), (5.61) and

$$\tau = \frac{\lambda}{\sqrt{\langle v^2 \rangle}}. \tag{7.49}$$

We find

$$\frac{1}{\lambda} = \frac{m^{*2} u^2 kT}{\pi \hbar^4 \rho c_L^2}. \tag{7.50}$$

The proper way to proceed in order to find the conductivity is to insert (7.47) or (7.48) into (7.7) or (7.8). Information which is semi-quantitative results, however, from insertion of (7.50) into (7.10) or (7.11) with the help of (7.49).

This yields for the mobility μ

$$\mu = \frac{\pi e \hbar^4 \rho c_L^2}{m^{*3} u^2 k T \sqrt{\langle v^2 \rangle}}.$$ (7.51)

Experimentally, the most easily verifiable part of (7.51) is the variation of μ with temperature. If we deal with a semiconductor having non-degenerate carriers then $\sqrt{\langle v^2 \rangle}$ is proportional to $\sqrt{T}$ and μ comes out to be proportional to $T^{-\frac{3}{2}}$. For metals, on the other hand, the mobility should vary as $1/T$ because the current is carried by electrons at the Fermi surface for which $\sqrt{\langle v^2 \rangle}$ has a large value independent of temperature. Both these predictions are verified, if the temperature is sufficiently high. At low temperature a double discrepancy arises: first the imperfections become the dominant scatterers, and second even if thermally excited lattice waves remain dominant their quantization must be considered. In some cases, such as p-type silicon and germanium, formula (7.51) is wrong even at high temperature. The reason seems to be the complicated band structure of these materials which suppresses a large part of the scattering by acoustical phonons; the optical branch of the vibration spectrum must then be included to yield a correct final formula.

PROBLEMS

1. Suppose that at an instant $t = 0$ a set of similar particles have a velocity distribution such that their speeds are all equal but their direction of motion is variable. Let the initial value of their mean drift velocity be $\langle v_0 \rangle$. Suppose that, from this moment on, they are subject to elastic scattering by stationary obstacles randomly distributed. Show that the drift velocity $\langle v \rangle$ relaxes in time according to the formula

$$\langle v \rangle = \langle v_0 \rangle \exp\left[-t \left\langle \frac{1 - \cos \theta}{\tau} \right\rangle \right].$$

Here θ is the angle of scattering by the obstacles and τ the free time between collisions. The pointed bracket indicates that the average of the quantity in brackets for all particles and collisions is to be taken.

2. Calculate the intrinsic conductivity of a piece of pure germanium at room temperature from the following data:
   (a) The constant of problem 1, Chapter 6, equals $5 \cdot 3 \cdot 10^{26}$ cm^{-6}.
   (b) The mobility of excess electrons equals 3800 cm^2/volt sec.
   (c) The mobility of defect electrons equals 1800 cm^2/volt sec.

3. Prove that electron-electron collisions make no contribution to the relaxation time controlling Ohm's law, unless they are of the 'Umklapp' type (see the discussion of equation (3.75)).

4. Derive from equation (7.32) the total power absorption arising from one spectral line, proceeding as for (7.33). Comment on the result.

5. Prove the statement made in the text that if the diagonal and off-diagonal matrix elements of the density matrix have different relaxation time then the line shape (7.32) is controlled by the off-diagonal relaxatior time.

6. Many derivations, and in particular (7.35), make use of the equivalence of the operators x and r for electrons confined to one band. Prove for a crystal having a centre of symmetry that the normalizing factor of the Bloch functions can be chosen in such a way that *all* matrix elements of x and r within the same band q are equal.

7. Derive (7.20), employing the methods of Chapter 6. Do not assume that $W(k)$ is necessarily quadratic in k.

BIBLIOGRAPHY

Almost all references of the last two chapters discuss electrical conductivity. In addition the following specific references are available.

KINETIC THEORY

S. CHAPMAN and T. G. COWLING. *The Mathematical Theory of Nonuniform Gases*, 2nd ed. (Cambridge, University Press, 1952.)

U. FANO. Description of states in quantum mechanics by density matrix and operator techniques. *Rev. Mod. Phys.* **29** (1957), 74.

ELECTRICAL CONDUCTION OF SOLIDS

J. BARDEEN. Electrical conductivity of metals. *J. Appl. Phys.* **11** (1940), 88–111.

F. J. BLATT. Theory of mobility of electrons in solids. *Solid State Physics*, **4** (1957), 199–366.

D. K. C. MACDONALD. Electrical conductivity of metals and alloys at low temperatures. *Handbuch der Physik*, **14** (1956), 137–97.

THEORY OF LINE BROADENING

C. J. GORTER. Paramagnetic Relaxation. (Amsterdam, Elsevier, 1947.)

C. H. TOWNES and A. L. SCHAWLOW. Microwave Spectroscopy, ch. 13. (New York, McGraw-Hill, 1955.)

There exists no modern summary on line broadening as it affects solids. A theoretical treatment containing the references to the most important original articles is: R. KUBO and K. TOMITA, *J. Phys. Soc. Japan*, **9** (1954), 888.

CHAPTER 8

SOLID COHESION AND CHEMICAL BONDING

The problem of solid cohesion, the 'why' of solids, is the ultimate touchstone of solid state theory. In order to appreciate its importance we may contrast solids with gases, for which molecular theory provides the 'why' of their existence and gives thereby also a rough understanding of their mechanical properties. From this point of view, solid state theory requires something transcending molecular theory. This something is not to be found in the previous chapters because all these chapters share the acceptance of the crystal as a matrix or, so to speak, as the theatre of the events described.

The answer to the 'why' of solids is not complicated in principle. Molecules aggregate into solids because of intermolecular attractive forces. If the temperature is low enough, so that these forces can overcome thermal agitation, then the molecules will prefer a condensed state to the gaseous phase. The condensed state realized among all possible ones is the one minimizing the free energy of the total system. This answer shows immediately that the motivation for solidification is not specific to the solid aggregate. The same forces can be examined under conditions which isolate the molecules or atoms. Only experience can answer the question which condition is most suitable for their elucidation. The answer has not been uniform. Crystals had an early lead in furnishing clues to the understanding of forces between atoms. This lead has gradually been lost. It is clear that if real precision work is wanted concerning the forces between two atoms, then it is reasonable to turn to conditions where these atoms are not being interfered with by the close presence of other bodies. For this reason the theorist treating solid cohesion is apt to forget his subject, and to shift his attention to diatomic molecules or scattering phenomena. This shift is inherent in the subject matter, since it is based on the identity of interatomic forces in binary and in larger aggregates.

The easiest case for understanding solid cohesion is provided in molecular crystals by the bond between chemically inert atoms or molecules. Such structures show a weak attraction for each other because the electrons in their orbits are virtually in phase; this produces a weak unsaturated dipole-dipole attraction which is known under the name of *Van der Waals force*. To a certain approximation the potential of this attractive force can be given the form†

$$U = -\frac{3}{2}\frac{I_1 I_2 \alpha_1 \alpha_2}{I_1 + I_2}\frac{1}{R^6}, \tag{8.1}$$

where I_1 and I_2 are the ionization potentials, α_1 and α_2 the polarizabilities of the two structures and R the distance between them. To this attractive force repulsive forces add at short distance. Since the latter have much shorter range, it is approximately correct to estimate the cohesive energy at the experimentally observed R with the help of (8.1) only. Summation over all pairs in a crystal is not a serious problem because $1/R^6$ drops rapidly with distance. Results for some molecular crystals are collected in Table 1. The agreement is sufficiently convincing to show that solid cohesion is not per se a hard problem but becomes difficult only when the identification of the forces is difficult.

Table 1. *Cohesive energies of molecular crystals as calculated from equation* (8·1)

(After Syrkin and Diatkina.)

	I (eV)	α (Å³)	Cohesive energy (eV per molecule)	
			calc.	exp.
Ne	21·4	0·40	0·017	0·026
Ar	15·6	1·66	0·078	0·088
Kr	13·9	2·54	0·134	0·122
N_2	15·6	1·74	0·070	0·081
O_2	13·0	1·57	0·064	0·089
Cl_2	18·2	4·60	0·312	0·322
CH_4	14·5	2·58	0·107	0·117

CLASSICAL THEORY OF IONIC CRYSTALS

All the solids of Table 1 above are gases at room temperature. The fascination of the theory of solid cohesion lies in the solids existing at room temperature and higher; in these cases the cohesive energy is orders of magnitude larger than those of Table 1 and comparable with the bond strength of molecules. Its study

† H. Margenau, *Rev. Mod. Phys.* **11** (1939), 1 (equation (10)).

throws therefore a great deal of light on the chemical bond in general. In this sense the rest of this chapter is also a discussion of the chemical bond.

Among all types of chemical bonding, the most clear-cut case arises between alkali atoms and halogen atoms. To understand it we must only accept the fact that these materials bond as singly charged ions. To illustrate this, data pertaining to these molecules are collected in Table 2. The first column gives the compound, the second the interatomic distance d, the third the work D in eV of separating them *into ions*, the fourth column the quantity Dd/e^2 with D in ergs, the fifth the electric dipole moment μ, the sixth the quantity μ/ed. The three fundamental quantities d, D, μ are all obtained independently so that we have here a real check of the ionic concept. If this concept were valid exactly then the quantities Dd/e^2 in column 4 and μ/ed in column 6 would be unity.

Table 2. *Data concerning the alkali halide molecules*

Compound	$d \times 10^8$ cm.	D (eV)	Dd/e^2	$\mu \times 10^{18}$	μ/ed
KF	2·55	5·8	1·03	7·3	0·60
KCl	2·79	4·92	0·95	6·3	0·47
KBr	2·94	4·64	0·95	—	—
KI	3·23	4·51	1·01	6·8	0·44
NaCl	2·51	5·54	0·96	—	—
NaBr	2·64	5·33	0·98	—	—
NaI	2·90	5·14	1·09	4·9	0·35

For the first of these quantities we have here a check so remarkable that there can be little doubt about the essential correctness of the view that these molecules consist of two ions bonded electrostatically. However, the precision is partly fortuitous as can be seen from the last column whose numbers should also be 1 on a strictly ionic picture. There are several reasons why the check on D comes out so much more closely than on μ. First, the ionic picture is least acceptable at the actual interatomic distance where a repulsive force of some sort must actually cancel the electric attraction; at larger distances this is not so much the case. Now μ refers to this actual distance, while D is an accumulated potential energy as the ions are brought together from large distances. A second reason is a partial cancellation of errors of opposite sign in D. In addition to the attractive $1/r^2$ force we have the short range repulsion of the ion cores already mentioned (sometimes idealized as 'hard sphere' repulsion, and tabulated as 'ionic radii'). However there is another attractive force, varying as $1/r^5$, due to the polarizability of the ions; that this enters is particularly evident from the iodides, whose empirical energy of dissociation is large, while their dipole moment is very low. Polarizability goes roughly as the volume of the ion and hence is biggest for iodine.

To sum up these observations for the alkali halide molecules, we may say that the chemical bond between the constituents is explained semiquantitatively by the assumption that the constituents are charged and held together by their Coulomb attraction. The check is particularly good for the binding energies, but this may be partly fortuitous. The check of the dipole moments is poorer because of the polarizability of the atoms, particularly if they are large. This latter difficulty should not bother us in the case of the alkali halide crystals because there each ion is embedded in a symmetric environment, so that polarizability effects vanish to a high order. On the other hand, the effect of core repulsion will remain, and hence it should be possible to make an even better theory of alkali halide crystals than of molecules, provided this core repulsion can be handled in some empirical way.

The work just indicated was carried out by Born and others prior to the advent of quantum mechanics.† It is of fundamental importance because it carries through for a special case the identification of chemical, cohesive and electrostatic energy. The basic calculation required for ionic crystals is to determine the total potential energy of a set of point centres of force, arranged on a regular lattice such as shown in figs. 1.36 and 1.37. The force consists mainly of the Coulomb force arising from their charges, but there is in addition, as pointed out above, a short range repulsive force which is to be treated empirically. The early workers took its potential in the form $A/(r_{ij})^n$ where r_{ij} is the distance of the ion i from the ion j, and n is an undetermined exponent, whose value came out to be in the neighbourhood of ten. Since the force is essentially due to interpenetration of electron clouds it is somewhat more natural to represent it by an exponential term, which exists

† M. Born, *Atomtheorie des festen Zustandes* (Teubner, Leipzig 1923).

only between nearest neighbours in the lattice. The formal expression for the total potential energy of the crystal thus takes the form

$$U = \tfrac{1}{2} \sum_{i \neq j} \pm \frac{e^2}{r_{ij}} + A_1 \sum_{\langle i,j \rangle} e^{-A_2 r_{ij}}; \qquad (8.2)$$

here the second summation goes only over direct neighbour pairs. For the verification of the theory there are three pieces of experimental information: the cohesive energy, the equilibrium lattice parameter and the compressibility. Thus the possibility of checking on the basic postulates is built into the theoretical ansatz (8.2) since it contains only two arbitrary constants.

To make (8.2) convenient for application it is well to remember that many of the most common salts crystallize in the cubic system. The magnitude of each term depends then only on one quantity, the lattice parameter a, and the actual dependence on that parameter follows from dimensional analysis. The first term has the dimension e^2/a, and for each ion has negative sign. Similarly, the second term, has the form $A\,e^{-a/\rho}$, with new undetermined constants A and ρ. The expression (8.2) thus takes the form

$$U = -NM\frac{e^2}{a} + A\,e^{-a/\rho}. \qquad (8.3)$$

Here N is the number of molecules in the crystal and M is a dimensionless positive constant of the order unity. If the crystal may be taken as infinite, then M is a mathematically defined number which can be calculated once for all by the prescription

$$M = \Sigma \pm \frac{1}{r_i}, \qquad (8.4)$$

where the summation goes over the distances r_i from the origin to the lattice points of a lattice of unit lattice parameter; the signs are to be picked in accordance with the sign of the Coulomb interaction in the particular lattice. The resultant number is called the *Madelung number* for that lattice. Its computation poses some problem because the sum (8.4) is only conditionally convergent; the order of the terms and the manner of summation is therefore important. We shall not discuss this mathematical problem here.† The

† See the literature list below. Also H. M. Evjen, *Phys. Rev.* **39** (1932), 675.

result for the two ionic crystal structures shown in figs. 1.36 and 1.37 are:

$$\text{NaCl structure} \quad M = 3 \cdot 4951, \tag{8.5a}$$

$$\text{CsCl structure} \quad M = 2 \cdot 0354. \tag{8.5b}$$

Since the theory has two undetermined parameters A and ρ we must now proceed to eliminate these with the help of the experimental information. First, if a is the equilibrium lattice parameter (at absolute zero of temperature) we must have

$$dU/da = 0, \tag{8.6}$$

which from (8.3), yields for U

$$U = -NM\frac{e^2}{a}\left(1 - \frac{\rho}{a}\right). \tag{8.7}$$

To eliminate the parameter ρ of the repulsive potential the bulk modulus B may be employed. It is defined by

$$B = \Omega \frac{d^2 U}{d\Omega^2}, \tag{8.8a}$$

where the total volume Ω is related to the previous parameters through

$$\Omega = Na^3/j, \tag{8.8b}$$

$$j = \text{number of molecules per cubic cell.} \tag{8.8c}$$

Carrying out the double differentiation on (8.3) and using (8.6) we get for B the following expression

$$B = \frac{jMe^2}{9a^4}\left(-2 + \frac{a}{\rho}\right). \tag{8.9}$$

If this equation is used to determine ρ, and if the value so obtained is inserted into (8.7), expressions for the cohesive energy result which can be compared with chemical information. This comparison is carried out in Table 3. There is a remarkable degree of agreement observed, even though systematic deviations are noticeable. Even without the repulsive second term in (8.3) or (8.7) there is already semi-quantitative agreement. The raw Coulomb energy calculation for NaCl, for instance, yields 8·9 eV for the cohesive energy.

Table 3. *Comparison of the cohesive energies of the alkali halides*

(After Born and Huang.)

Substance	Structure	a (Å)	ρ (Å)	Cohesive energy eV per molecule	
				theor.	exp.
LiCl	NaCl	5·144	0·664	8·518	8·74
LiBr	NaCl	5·490	0·696	8·000	8·31
LiI	NaCl	6·000	0·748	7·337	7·81
NaCl	NaCl	5·628	0·656	7·895	8·02
NaBr	NaCl	5·962	0·666	7·494	7·63
NaI	NaCl	6·462	0·726	6·914	7·21
KCl	NaCl	6·278	0·648	7·190	7·28
KBr	NaCl	6·586	0·668	6·869	7·00
KI	NaCl	7·052	0·698	7·429	6·63
RbCl	NaCl	6·540	0·676	6·902	7·10
RbBr	NaCl	6·854	0·700	6·592	6·90
RbI	NaCl	7·326	0·704	6·208	6·50
CsCl	CsCl	4·108	0·368	6·490	6·84
CsBr	CsCl	4·287	0·383	6·221	6·61
CsI	CsCl	4·56	0·393	5·868	6·31

THEORY OF THE COVALENT BOND

The classical theory of ionic lattices constituted in its day a tremendous advance in the understanding of chemical and solid bonding. We know now, however, that this bond is only one of a number of types, and exists in its pure form only for salt-like compounds between elements of opposite electric polarity. When

Fig. 8.1. Electrostatic nature of the covalent bond; one electron bonding two positive ions.

chemical bonding occurs between atoms which are similar or identical then both atoms become positively charged and the valence electrons concentrate along the line joining them as shown in fig. 8.1; this is the *covalent bond*. This bond is much less accessible to a priori theoretical computation than the ionic bond, because the concentration of electrons is governed by many-body quantum mechanics and varies from case to case. The study must proceed therefore by prototypes, followed by qualitative generalizations. The prototype most easily amenable to calculation is the hydrogen molecular ion H_2^+. By the general idea underlying our treatment of solids (Born–Oppenheimer approximation), this system may be thought of to a first approximation as a one-body problem: one electron moving in the stationary field of two protons. To treat it, let the proton-proton distance have the fixed value R and let the electron be instantaneously located at P. Let the distances of P from the two protons be r_1 and r_2 (fig. 8.2). The electron then obeys the Schrödinger equation

$$-\frac{\hbar^2}{2m} \nabla^2 \psi - \left(\frac{e^2}{r_1} + \frac{e^2}{r_2}\right) \psi = E(R)\,\psi. \qquad (8.10)$$

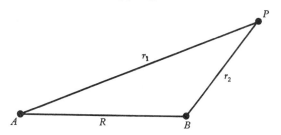

Fig. 8.2. Parameters for the molecular hydrogen ion problem.

This equation can be solved in elliptic co-ordinates with the two proton positions A and B as focal points. In other words we set

$$\xi = \frac{r_1 + r_2}{R} \quad (1 \leqq \xi < \infty), \qquad (8.11a)$$

$$\eta = \frac{r_1 - r_2}{R} \quad (-1 \leqq \eta \leqq +1), \qquad (8.11b)$$

and use as the third co-ordinate the azimuth ϕ of the electron about the line AB. The equation (8.10) then takes the form

$$-\frac{2\hbar^2}{mR^2}\left[\frac{1}{\xi^2-\eta^2}\frac{\partial}{\partial\xi}\left\{(\xi^2-1)\frac{\partial\psi}{\partial\xi}\right\} + \frac{1}{\xi^2-\eta^2}\frac{\partial}{\partial\eta}\left\{(1-\eta^2)\frac{\partial\psi}{\partial\eta}\right\}\right.$$
$$\left. + \frac{1}{(\xi^2-1)(1-\eta^2)}\frac{\partial^2\psi}{\partial\phi^2}\right] - \frac{4e^2\xi}{R(\xi^2-\eta^2)}\,\psi = E(R)\,\psi. \qquad (8.12)$$

Since ϕ does not appear explicitly in the equation angular momentum along the AB axis is conserved: in general ψ will contain a factor $e^{im\phi}$ and, in the ground state, will presumably be independent of ϕ. For the variables ξ and η we can use the method of separation of variables, setting

$$\psi(\xi, \eta) = X(\xi) \times Y(\eta) \qquad (8.13)$$

and substitute this into (8.12). We get the two ordinary differential equations

$$\frac{d}{d\xi}\left\{(\xi^2 - 1)\frac{dX}{d\xi}\right\} + \left(\frac{mR^2E}{2\hbar^2}\xi^2 + \frac{2mRe^2}{\hbar^2}\xi + A\right) X = 0, \quad (8.14)$$

$$\frac{d}{d\eta}\left\{(\eta^2 - 1)\frac{dY}{d\eta}\right\} + \left(\frac{mR^2E}{2\hbar^2}\eta^2 + A\right) Y = 0. \qquad (8.15)$$

The boundary conditions (8.11) at the end of the intervals determine eigenvalues for the energy E and the separation constant A for each value of R. No procedure is known at present to do this analytically.

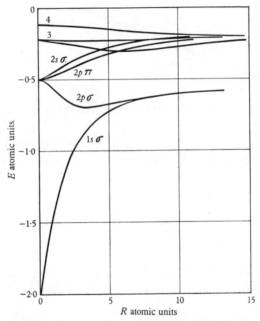

Fig. 8.3. Plot of the electronic energy levels of H_2^+ versus the internuclear distance (after Bethe).

The result for E is therefore shown in the form of a graph in fig. 8.3. To obtain the effective potential for the motion of the protons we must add to E the repulsive potential e^2/R between them:

$$U(R) = E(R) + \frac{e^2}{R}. \qquad (8.16)$$

Curves for $U(R)$ are shown in fig. 8.4. The lowest one is of importance to us. Its stable minimum occurs at

$$R_0 = 2 \cdot 00 \frac{\hbar^2}{me^2} = 1 \cdot 06 \, \text{Å}, \qquad (8.17a)$$

$$U_0 = -0 \cdot 603 \frac{me^4}{\hbar^2} = -16 \cdot 3 \, \text{eV}. \qquad (8.17b)$$

Subtraction from $|U_0|$ of one Rydberg (which is the energy of a hydrogen ion and atom at infinite separation) yields the energy D of dissociation of the molecular ion:

$$D = 0 \cdot 103 \frac{me^4}{\hbar^2} = 2 \cdot 8 \, \text{eV}. \qquad (8.17c)$$

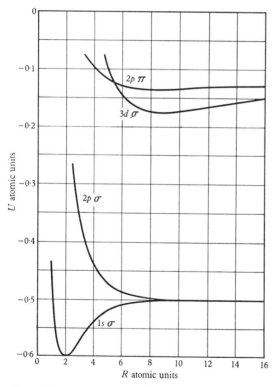

Fig. 8.4. Plot of the total energy of H_2^+ versus the internuclear distance; the molecular ion is stable, because the lowest curve has a minimum when R equals two atomic units (after Bethe).

The equations (8.17) are important, because we have here a result of an exact calculation providing us with a covalently bonded molecular system. The numerical result is in fact close to the upper limit of the covalent bond energy which can be provided by one electron between two atom cores. Qualitatively, we can understand this result as arising from the electrostatic attraction which is produced when a negative charge is placed between two positive charges (fig. 8.1). This attraction would already exist in classical physics. Specifically quantum mechanical reasoning is needed, not for this, but to find out how far the arrangement of fig. 8.1 can

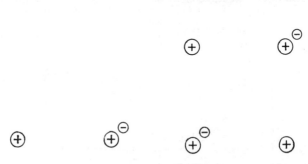

Fig. 8.5. Antibonding configuration of an electron and two protons.

Fig. 8.6. Superposition of states as applied to H_2^+; the molecular ion is considered a superposition of two configurations, each of which consists of an atom and an atomic ion.

be compressed in scale before the electron, with its finite wave length, escapes from its energetically favourable position into an arrangement as shown in fig. 8.5.

Historically, the theory of Heitler and London concerning the hydrogen molecule has preceded the calculations given above for the hydrogen molecular ion. Their pioneering work introduced into molecular theory the fruitful notion of treating the molecule not as a system to be studied ab initio, but as a linear superposition of atomic structures which are termed *participating states*. This viewpoint is inspired partly by chemical reasoning, and partly also by the great simplifications which result from the possibility of dealing only with atomic wave functions or *atomic orbitals*. In H_2^+ the atomic orbital method can be applied only in a single way, namely by considering H_2^+ made up out of a superposition of the two configurations of fig. 8.6. In other words, we write

$$\Psi = \psi(\mathbf{x} - \tfrac{1}{2}\mathbf{R}) + \psi(\mathbf{x} + \tfrac{1}{2}\mathbf{R}) \tag{8.18}$$

or

$$\Psi = \psi(\mathbf{x} - \tfrac{1}{2}\mathbf{R}) - \psi(\mathbf{x} + \tfrac{1}{2}\mathbf{R}) \tag{8.19}$$

for the wave function of the electron. Here $\psi(\mathbf{x})$ is the ground-state wave function of an electron in the hydrogen atom and $\mathbf{R}$ is

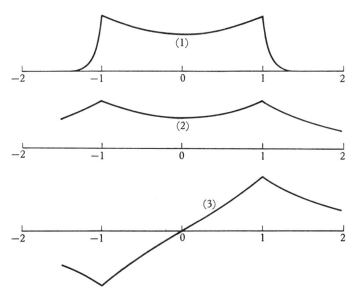

Fig. 8.7. Ground-state wave functions of H_2^+, plotted along the line of centres; exact wave function (1), bonding combination of atomic functions (2), antibonding combination of atomic functions (3).

the interatomic distance vector. The two combinations (8.18) and (8.19) are prescribed for us by the reflection symmetry of the system about the midpoint between the two protons.

A straightforward way of comparing (8.18) and (8.19) with the exact solutions of (8.14) and (8.15) is to compare the wave functions to which they lead. This is done in fig. 8.7, where the true ground-state wave function and the two approximations are plotted along the line of centres for comparison. The similarity of the first two is evident, as well as the dissimilarity of the third. In fact while both the first two wave functions show the preferential position of the electron between the protons, the third shows a preferential location outside. On electrostatic grounds we must, therefore, expect the third wave function to lead to repulsion of the two protons. The combination (8.19) is therefore termed the *antibonding* combination of wave functions and (8.18) the *bonding* one. Fig. 8.8 shows for the same three wave functions the energy, defined as the expectation value of the total hamiltonian, as a function of R. The difference between the energies 3 and 2 arises

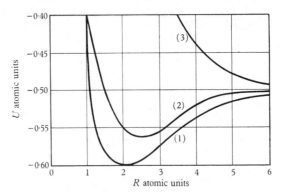

Fig. 8.8. Plot of energy versus internuclear distance
for the three wave functions of fig. 8.7.

from the product of the two wave functions entering into (8.18) and (8.19); it may therefore be said to originate from the coupling of the two atomic systems. In analogy to the frequency shift arising from the coupling of oscillators, the energy shift is therefore often referred to as the *resonance energy* of the two participating states; similarly the phenomenon of the stable minimum in curve 2 is called *resonance stabilization*. Actually, the stable minimum in curve 1 is only very roughly described by the approximation 2.

We can predict, on general grounds, that the H_2^+ system, although stable against dissociation into a proton and a hydrogen atom, is not stable against other chemical changes. In the first place, the system is positively charged which might predispose it somewhat toward accepting another electron. In the second place, we have just studied a low-energy orbital provided for an electron by the framework of two protons. This orbital, so far, contains only one electron. Now we know from atomic theory, that because of electron spin, every orbital is capable of holding two electrons. Hence the process of bringing in a second electron and placing it with opposite spin on the same bonding orbital should proceed with energy gain. We also know from atomic theory that the repulsion of the two electrons, while modifying the wave functions somewhat, is not sufficient to invalidate this reasoning. Indeed, the experiment shows that 15·4 eV is gained by the addition of a second electron. This differs very little from the value 16·3 for the first electron. This figure also increases the energy of dissociation as compared to that of the molecular ion, namely with (8·17)

$$D_{H_2} = 2\cdot8 + 15\cdot4 - 13\cdot5 = 4\cdot7 \, eV. \qquad (8.20)$$

If we keep this figure in mind, it will not surprise us too much to find that the electron-pair bond is the most common bond of chemistry. For we have just seen its extra stability over the one-electron bond; furthermore, if a third electron is brought into play then we have to find for it, because of the Pauli principle, a ψ-function orthogonal to the one used so far. In the present case, this is quite impossible to do with gain in bond energy; however, for larger atoms it is possible as will be discussed below.

The tendency of compounds to contain an even number of electrons is a very strong one in chemistry and finds its explanation in the approximate existence of electron orbitals, the electron spin and the Pauli principle. On the other hand, there are no features discernible in the electron-pair bond which are not already present in the one-electron bond. Indeed, the value (8.20) is considerably less than the double of (8.17c). This means that *per bonding electron*, the one-electron bond is the strongest one. It seems therefore safe to assume that the multi-electron bonds are simply enfeebled replicas of it, and that no new physical idea has to be brought in to 'explain' them.

If we now pass to a quantitative theoretical discussion of H_2, then we must admit immediately that no rigorous calculation of H_2 has ever been made, and all results are based on approximations. We shall discuss here only the approximations based on the principle of superposition of states. The principle was, in fact, invented for this system by *Heitler and London*,† who postulated the two participating arrangements shown in fig. 8.9. Each arrangement consists of two undisturbed hydrogen atoms. They differ only by having the electrons switched in one system compared to the other. In other words, if $\mathbf{x}_1$ and $\mathbf{x}_2$ are the position vectors of the two

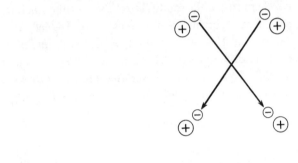

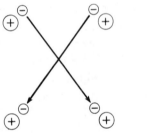

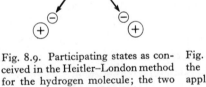

Fig. 8.9. Participating states as con- Fig. 8.10. Generalized conception of ceived in the Heitler–London method the method of participating states as for the hydrogen molecule; the two applied to the hydrogen molecule. states are not physically distinct.

electrons and $\mathbf{R}$ the internuclear distance vector, the Heitler–London two-electron wave function reads

$$\Psi = \psi(\mathbf{x}_1 - \tfrac{1}{2}\mathbf{R})\,\psi(\mathbf{x}_2 + \tfrac{1}{2}\mathbf{R}) + \psi(\mathbf{x}_2 - \tfrac{1}{2}\mathbf{R})\,\psi(\mathbf{x}_1 + \tfrac{1}{2}\mathbf{R}), \quad (8.21)$$

where ψ is the $1s$ eigenfunction of the hydrogen atom. This is the bonding wave function. A second, antibonding wave function can be constructed from the same states, namely

$$\Psi = \psi(\mathbf{x}_1 - \tfrac{1}{2}\mathbf{R})\,\psi(\mathbf{x}_2 + \tfrac{1}{2}\mathbf{R}) - \psi(\mathbf{x}_2 - \tfrac{1}{2}\mathbf{R})\,\psi(\mathbf{x}_1 + \tfrac{1}{2}\mathbf{R}). \quad (8.22)$$

† W. Heitler and F. London, *Zeits. f. Phys.* **44** (1927), 455.

(8.21) and (8.22) behave qualitatively the same way as (8.18) and (8.19). The distribution of charge, and hence the bonding, is controlled by Ψ^2; it differs, therefore, from the atomic charge distribution by the double product term. This charge shift is into the region between the two protons in (8.21) and away from that region in (8.22).

There is a second way in which the basic idea of participating atomic states can be employed for constructing molecular wave functions. We can generalize the idea of fig. 8.6 to H_2, not by combining the wave functions for the two atoms as a whole, as in (8.21) and (8.22), but by combining one-electron wave functions from the two atoms to produce a *molecular orbital* wave function, into which two electrons of opposite spin can be placed. In the case of H_2 this means simply that we take the wave function (8.18) for either electron. The total two-electron wave function (without spin) reads then

$$\Psi = \{\psi(\mathbf{x}_1 - \tfrac{1}{2}\mathbf{R}) + \psi(\mathbf{x}_1 + \tfrac{1}{2}\mathbf{R})\}$$
$$\times \{\psi(\mathbf{x}_2 - \tfrac{1}{2}\mathbf{R}) + \psi(\mathbf{x}_2 + \tfrac{1}{2}\mathbf{R})\}. \quad (8.23)$$

When this wave function is multiplied out it is seen that it ascribes equal weight to the four participating states of fig. 8.10. In reality, the two ionic states are participating with a very much smaller coefficient, and the wave function (8.21) is somewhat better for H_2 than (8.23). For most other simple molecular systems, the reverse is found to apply, however; the appropriate generalization of (8.23) is found to give lower energy values than the best generalization proposed of (8.21). The great flexibility of products of one-particle wave functions may be partly responsible for this. A product of 'tight binding' wave functions of the form (5.84) represents, for instance, a straightforward generalization to a crystal of the concept contained in (8.23). No such easy generalization of (8.21) has been worked out up to now. The relative merit of the two viewpoints for solids is still a controversial question. Results for molecules from these and other approximations are clearly outside the scope of this book.†

Through (8.17), the existence of the covalent bond has been proved directly from quantum mechanics for H_2^+. Calculations for

† See the literature list at the end of the chapter for further information.

H_2 and its two-electron bond have been sufficiently numerous to leave little doubt concerning the nature of that molecule. Beyond this, calculations from first principle rapidly become too complicated to be feasible in practice, and therefore simplifications are needed. The first simplification is the adoption of one-electron orbitals. From the standpoint of these one-electron orbitals the widespread existence of a saturated two-electron bond becomes plausible; for the Pauli principle and spin combine to preclude the placement of a third electron into a bonding orbital occupied by two others. Already this reasoning is, however, partly backward from experience. This inductive aspect of theoretical chemical reasoning is very obvious in the discussion of all molecules more complicated than H_2. In particular, the method of participating states is a systematization of chemical experience. Actually, there is nothing in the general theory demanding its adoption, and its results in the simple cases just discussed are only mediocre. However, the method has the advantage of making computational procedures possible in chemistry, and of systematizing a fair amount of empirical information. We shall show this by deriving with its help the concept of a *valence* associated with the position of an atom in the periodic table, and by accounting for the phenomena of *directed valence* and *multiple bonds*.

In ionic compounds, valence is associated in an obvious way with the number of electrons an atom must either donate or accept so that its electron shell forms a noble gas configuration. In covalent bonding, the picture is that a bond is formed by two electrons, and that these electrons must be counted as forming a constituent part of either atom. This rule is reproduced in the method of participating states, either in the form (8.21), or in the form (8.23). According to this method, no covalent-bond wave function can be constructed without 'using up' an atomic orbital on either atom to be bonded, and saturation arises when orbitals beyond a closed shell would have to be employed for bond formation. In this way the method explains the standard valencies of nitrogen, oxygen and the halogens. It also accounts for some details; it shows, for instance, that hydrogen is covalently monovalent while the alkali metals are not.

A second notion explained by this method is the phenomenon

of directed valence. The bond formed by $1s$ electrons saturates after one linkage because the K-shell can only accept two electrons. Already in the next shell, from Li on, more flexibility arises because the shell can hold eight electrons, two s's and six p's. From this base several potential bonding wave functions can be constructed which reach out in different directions. Such mixing of atomic wave functions is called *hybridization*; it becomes allowed as soon as the atoms cease to be in a spherically symmetric environment. The simplest hybridization arises from the mixture of one s and one p orbital to construct two bonding orbitals pointing in opposite directions, namely

$$\left. \begin{aligned} \Psi_+ &= \psi_s(r) + \psi_p(r)\cos\vartheta = \psi_s(r) + \frac{1}{r}\psi_p(r)\,z, \\ \Psi_- &= \psi_s(r) - \psi_p(r)\cos\vartheta = \psi_s(r) - \frac{1}{r}\psi_p(r)\,z. \end{aligned} \right\} \tag{8.24}$$

By combining two p-orbitals with an s-orbital we can make three bonds pointing at $120°$ in a plane:

$$\left. \begin{aligned} \Psi_1 &= \psi_s(r) + \frac{1}{r}\psi_p(r)\,x, \\ \Psi_2 &= \psi_s(r) + \frac{1}{r}\psi_p(r)\left(-\tfrac{1}{2}x + \frac{\sqrt{3}}{2}y\right), \\ \Psi_3 &= \psi_s(r) + \frac{1}{r}\psi_p(r)\left(-\tfrac{1}{2}x - \frac{\sqrt{3}}{2}y\right). \end{aligned} \right\} \tag{8.25}$$

Finally, by the use of all three p-functions and an s-function, bonds pointing in the four directions of a regular tetrahedron are obtained:

$$\left. \begin{aligned} \Psi_a &= \psi_s(r) + \frac{1}{r}\psi_p(r)\,(x+y+z), \\ \Psi_b &= \psi_s(r) + \frac{1}{r}\psi_p(r)\,(x-y-z), \\ \Psi_c &= \psi_s(r) + \frac{1}{r}\psi_p(r)\,(-x+y-z), \\ \Psi_d &= \psi_s(r) + \frac{1}{r}\psi_p(r)\,(-x-y+z). \end{aligned} \right\} \tag{8.26}$$

The advantage of hybridized orbitals lies in the strong localization of the electrons along a narrow cone of directions from the parent atom. To the extent that this concentration is produced by interference of waves of equal wave length the kinetic energy of the electron is not appreciably raised; yet at the same time the potential energy is diminished because the electrostatic ideal of fig. 8.1 is more perfectly realized than for non-hybridized orbitals. Thus extra stability is gained for the molecule or crystal as a whole by having a definite angular arrangement imposed upon the bonded atoms. The bonding system (8.26) is of particular importance for

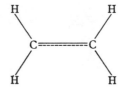

Fig. 8.11. The molecule of ethylene.

elements of the fourth group of the periodic table (C, Si, Ge), since they have just the right number of electrons to actually form all four covalent bonds. In the lower lines of the periodic table more complicated geometrical arrangements of bonds become possible with the help of d wave functions.

A third notion follows naturally from the principle of superposition of atomic states: the possibility of bonds involving more than two electrons. This arises again from the presence of several orbitals in the same shell. The simplest example is ethylene C_2H_4 shown in fig. 8.11. The molecule is planar and may be thought of as using ten of its twelve outer shell electrons to form bonds according to the scheme of equation (8.25). This leaves over the wave function $(z/r)\,\psi_p(r)$ which, at each carbon atom, sticks out of the plane of the paper with opposite sign on the two sides. Because of this the wave function is automatically orthogonal to the C—C bonding function. Since only two electrons have to be accommodated, it is energetically advantageous to make from them the combinations (8.18) and (8.19) and to place both electrons in the former. This is the so-called π-bond which, together with the original bond (called

σ-bond) forms the double bond of carbon. The argument implies that the C=C bond has less than the double strength of C—C, partly for the reasons given for H_2 as against H_2^+, namely, that the addition of subsequent electrons to a bond increases the bond strength less than proportionally because of electron-electron repulsion; a second reason for this arises specifically for π-bond electrons: they are forced, because of the orthogonality condition, to have zero density in the plane of the molecule where they would be most efficient for bonding. This relative weakness of the double bond is indeed verified, the bonding energies of the double and single bond in carbon being 4·9 eV and 3·6 eV respectively.

A fourth notion which comes out of the generalized theory of valence is the idea of the *electron-deficient bond*. In a sense, the hydrogen molecular ion has an electron deficient bond; the sense we want to impart to it here arises when there is opportunity to form a larger number of bonding orbitals than there are electrons to fill them. This scarcity of electrons leads to a state of affairs which is the very opposite of saturated valence. For low atomic number the scarcity is confined to the left-hand side of the periodic table, but for high atomic number it occurs all over.

Generally, the electron deficiency of this type of bond is so severe as to make the substances metals, that is materials in which the bonding electrons are free to move in an applied electric field, and for which the entire crystal is essentially one molecule. A more limited type of deficiency arises in beryllium and boron. Saturation can be reached for those two atoms by the relatively small step of somehow arranging for bonds to four neighbours in the manner indicated by the wave functions (8.26), even though the available number of bonding electron pairs is only three. The classical example is boron hydride B_2H_6. In order to reach fourfold co-ordination at the boron atom two of its bonding electrons are made to service simultaneously two B—H bonds as shown in fig. 8.12. Another example is $Be(CH_3)_2$ shown in fig. 8.13. There the basic unit is already no longer a simple molecule but a chain. It is in the nature of this type of bond that a second electron wave function can be placed in the B^{H}B space which is also bonding in this instance: the lowest wave function is symmetric about the central vertical plane, but an antisymmetric one would still have correct location for bonding. In this instance this higher function is sufficiently separated to be inaccessible to the electron. As soon as the wave functions become so crowded in energy space as to permit free passage of thermal electrons, the metallic state is reached.

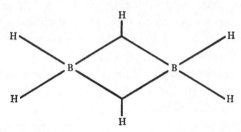

Fig. 8.12. The molecule of boron hydride; the substance is electron-deficient for normal tetrahedral bonding about a boron atom; it uses therefore one electron pair to bond two borons and a hydrogen.

The covalent bond, in its saturated form, co-operates with other forms of bonding in holding solids together. Very rarely does it act alone. The most important case of a crystal which is entirely covalently bonded is diamond and its homologues in the fourth column of the periodic table. The structure, discussed on p. 33,

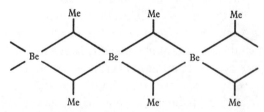

Fig. 8.13. Schematic representation of the crystal of dimethyl beryllium; there is an electron-deficient bond connecting one methyl group simultaneously to two berylliums.†

shows, indeed, each carbon in fourfold co-ordination to four other carbons, according to the tetrahedral bonding scheme (8.26). That we are dealing here with saturated valence is evident from the open structure of the crystal, which cannot be understood otherwise. A second argument is that the cohesive energy of diamond agrees closely with the 'bond energy' of the C—C bond deduced

† R. E. Rundle, *J. Chem. Phys.* 18 (1950), 1125.

from organic chemistry.† There are a number of other crystals which belong more or less in this category because of their fourfold co-ordination, notably SiC and ZnS. In the latter case, a certain amount of ionic character is probably already mixed in. This is even more so for the oxides.

THEORY OF THE COHESIVE ENERGY OF METALS

The electron-deficient covalent bond becomes, in crystals, the metallic bond. Metals are recognized by the metallic type of electrical conductivity (large conductivity with negative temperature coefficient) and their close packed, or nearly close packed, lattice structure. Their most common crystalline forms are face-centred cubic, hexagonal close packed, and body-centred cubic. These forms were discussed in Chapter 1. This type of crystal structure predominates in the solid form of the pure elements. When a mixture of two kinds of atoms is bonded in this way then it is often not at all specific about the proportions in which the constituents occur; instead of a compound we have then an *alloy*. This should not blind us to the fact that an alloy is held together by the same chemical forces as respectable compounds; we have to remember only that no specific valences can be asigned to the constituents, because their bonds are electron deficient.

When it comes to applications to specific metals rather than qualitative discussion, the method of superposition of atomic states does not impose itself naturally here as it did for saturated bonds. Its specific valence aspects are washed out by an oversupply of atomic wave functions among which the electrons must be thought of as resonating. Superpositions of the type (8.18) or (8.19), yielding one-electron crystal orbitals, are occasionally used in solid state work; they are the 'tight binding' wave functions introduced in (5.84). However, having arrived at a non-localized picture for metallic valence electrons, starting from the idea of localized states, we do better to discard these crutches and to build our non-localized electronic wave functions from the ground up. This is what has already happened essentially in Chapter 5. Since the

† The C—C bond strength is about 85 kcal. per mole in diamond and in ethane. Table 10–1 in Pauling's *Nature of the Chemical Bond* contains an error at this point.

picture developed there was found useful in explaining other properties of metals, there is no reason why it should not yield the cohesive energy also. Such calculations have been undertaken and will now be described.

The Schrödinger equation of quantum mechanics can be thought of as the result of a variational principle. Suppose we have a system of particles whose motion is governed by a hamiltonian $\mathcal{H}$. Then it may be shown that the stationary states of the system result from the formula

$$\delta E = \delta \left[\frac{\int \Phi \mathcal{H} \Psi \, d\tau}{\int \Phi \Psi \, d\tau} \right] = 0, \tag{8.27}$$

where the variation is taken independently with respect to the wave functions Φ and Ψ. When this variation is taken, Φ and Ψ are independently found to obey the Schrödinger equation, and E comes out to be the energy eigenvalue of those functions. The principle has found a widespread application in atomic theory because it permits the insertion of trial functions into (8.27). With such functions an approximate value of the energy may be obtained, and the best possible choice among those trial functions is singled out. The many atomic applications of this method are beyond the scope of this book. In solid state physics the principle is used primarily to supplement the one-electron theory of solids. One assumes for the valence electrons a wave function of the type postulated in Chapter 5:

$$\Psi = \prod_{i=1}^{N} u_i(\mathbf{x}_i), \tag{8.28a}$$

$$\Phi = \prod_{i=1}^{N} v_i(\mathbf{x}_i); \tag{8.28b}$$

Then one inserts this type of function into (8.27) and forms the variation. The result is a repetition of the same equation for each u_i and v_i, the so-called *Hartree* or *self-consistent field equation.*

In the form in which the Hartree equation is applied to solids no attempt is made to include the inner electrons as such. Instead, they, together with the nuclei, are said to produce an ion core potential $V(\mathbf{x})$ in which the valence electrons are moving. Clearly this potential has the periodicity of the lattice, that is, it obeys (5.45). One thus ends up with a hamiltonian for the valence electrons only, which has the form

$$\mathcal{H} = \sum_i \left(\frac{p_i^2}{2m} + V(\mathbf{x}_i) \right) + \sum_{i<j} \frac{e^2}{|\mathbf{x}_i - \mathbf{x}_j|}, \qquad (8.29)$$

where the summations extend over all valence electrons.

Equations (8.29) and (8.28) are now inserted into (8.27) yielding

$$\delta E = \delta \left[\sum_i \frac{\int v_i(\mathbf{x}) \left(\frac{p^2}{2m} + V(\mathbf{x}) \right) u_i(\mathbf{x}) \, d\mathbf{x}}{\int v_i(\mathbf{x}) u_i(\mathbf{x}) \, d\mathbf{x}} \right.$$

$$\left. + \sum_{i<j} \frac{\iint v_i(\mathbf{x}) v_j(\mathbf{y}) \frac{e^2}{|\mathbf{x}-\mathbf{y}|} u_i(\mathbf{x}) u_j(\mathbf{y}) \, d\mathbf{x} \, d\mathbf{y}}{\int v_i(\mathbf{x}) u_i(\mathbf{x}) \, d\mathbf{x} \int v_j(\mathbf{y}) u_j(\mathbf{y}) \, d\mathbf{y}} \right]. \qquad (8.30)$$

The variation can be taken independently with respect to each function; the variation with respect to v_i yields

$$\int \delta v_i(\mathbf{x}) \left[\left(\frac{p^2}{2m} + V(\mathbf{x}) \right) u_i(\mathbf{x}) \right.$$

$$\left. + \sum_{j \neq i} \frac{\left(\int v_j(\mathbf{y}) \frac{e^2}{|\mathbf{x}-\mathbf{y}|} u_j(\mathbf{y}) \, d\mathbf{y} \right) u_i(\mathbf{x})}{\int v_j(\mathbf{y}) u_j(\mathbf{y}) \, d\mathbf{y}} - \epsilon_i u_i(\mathbf{x}) \right] d\mathbf{x} = 0, \qquad (8.31)$$

where $\epsilon_i = \dfrac{\int v_i(\mathbf{x}) \left(\frac{p^2}{2m} + V(\mathbf{x}) \right) u_i(\mathbf{x}) \, d\mathbf{x}}{\int v_i(\mathbf{x}) u_i(\mathbf{x}) \, d\mathbf{x}}$

$$+ \sum_{j \neq i} \frac{\iint v_i(\mathbf{x}) v_j(\mathbf{y}) \frac{e^2}{|\mathbf{x}-\mathbf{y}|} u_i(\mathbf{x}) u_j(\mathbf{y}) \, d\mathbf{x} \, d\mathbf{y}}{\int v_i(\mathbf{x}) u_i(\mathbf{x}) \, d\mathbf{x} \int v_j(\mathbf{y}) u_j(\mathbf{y}) \, d\mathbf{y}}. \qquad (8.32)$$

ϵ_i is the same for the u's and the v's and, after some integrations by parts, the variational equations (8.31) are also the same. By the

17   W S T

general idea of variational calculus, the square brackets of (8.31) must be zero, since they are multiplied by arbitrary variations. To show the relationship of v_i to u_i, assume a small degeneracy-lifting field added to the hamiltonian; if the field is electric it seems to allow $v_i(\mathbf{x}) = u_i(\mathbf{x})$ rather than $v_i = u_i^*$; this is all right in this case because the wave functions can be made real. The reader is invited to check for himself that a small magnetic field yields in (8.31) conjugate complex equations for the two functions, so that the proper relationship is

$$v_i(\mathbf{x}) = u_i^*(\mathbf{x}), \qquad (8.33)$$

which yields finally the Hartree self-consistent field equations:

$$-\frac{\hbar^2}{2m} \nabla^2 u_i(\mathbf{x}) + V(\mathbf{x}) u_i(\mathbf{x}) - \epsilon_i u_i(\mathbf{x})$$

$$+ \sum_{j \neq i} \frac{\int u_j^*(\mathbf{y}) \dfrac{e^2}{|\mathbf{x}-\mathbf{y}|} u_j(\mathbf{y}) \, d\mathbf{y}}{\int u_j^*(\mathbf{y}) u_j(\mathbf{y}) \, d\mathbf{y}} u_i(\mathbf{x}) = 0 \quad (8.34)$$

with an ϵ_i given by substitution of (8.33) into (8.32). Actually, this value of ϵ_i is implicitly contained in (8.34): it results by multiplication with u_i^* and integration; in other words, the energy eigenvalue equals its expectation value.

There is a clear-cut physical interpretation of (8.34). A particular electron i is seen to have an effective one-particle hamiltonian which consists of the parts specific to it plus an average potential resulting from the other electrons, which are treated as charge clouds distributed according to their own wave functions. Hence the name self-consistent field method.

An interesting complication arises when the energy is computed as the expectation value of the total hamiltonian (8.29). This energy is not the sum of the energies ϵ_i; for we see from (8.32) that this would count the electron-electron interaction twice. The true expression is

$$E = \sum_i \epsilon_i - \sum_{i<j} \frac{\int u_i^*(\mathbf{x}) u_j^*(\mathbf{y}) \dfrac{e^2}{|\mathbf{x}-\mathbf{y}|} u_i(\mathbf{x}) u_j(\mathbf{y}) \, d\mathbf{x} \, d\mathbf{y}}{\int u_i^*(\mathbf{x}) u_i(\mathbf{x}) \, d\mathbf{x} \int u_j^*(\mathbf{y}) u_j(\mathbf{y}) \, d\mathbf{y}}. \qquad (8.35)$$

The explanation of this phenomenon is that if the energies are referred to a state in which all electrons are scattered at infinity then $-\epsilon_i$ is the energy needed to remove the electron i alone from the crystal while $-E$ equals the energy needed to remove them all. Now as the electrons are removed one by one their ionization potentials increase; this increase is measured by the second term in (8.35).

A relatively straightforward improvement of the Hartree method should be mentioned here. The trial solution (8.28) does not take into account the Pauli exclusion principle. One can take care of this in a superficial way by seeing to it that not more than two electrons are placed into any one eigenstate. The proper way is, however, to make the total wave function antisymmetric. Now the trial form (8.28) permits the generalization (5.87) which takes into account the Pauli principle for valence electrons of equal spin. This function reads

$$\Phi = \begin{vmatrix} u_1(x_1)u_1(x_2) & \cdots & u_1(x_{\frac{1}{2}N}) \\ u_2(x_1) & & \\ & \ddots & \\ & & u_{\frac{1}{2}N}(x_{\frac{1}{2}N}) \end{vmatrix} \times \begin{vmatrix} u_1(x_{\frac{1}{2}N+1}) & \cdots & u_1(x_N) \\ & \ddots & \\ & & u_{\frac{1}{2}N}(x_N) \end{vmatrix}$$

$$\times \alpha(1)\alpha(2)\ldots\alpha(\tfrac{1}{2}N)\beta(\tfrac{1}{2}N+1)\beta(\tfrac{1}{2}N+2)\ldots\beta(N). \quad (8.36)$$

With this new wave function one goes through the variational procedure discussed previously with (8.28) as trial function. The result of this procedure are the so-called *Fock equations*, whose derivation, very similar to the preceding one, will not be given here. They differ from the Hartree equations only in the term giving the electron-electron interaction. If two electrons i and j have equal spin then the electron i at the position $\mathbf{x}$ is influenced by the electron j at the position $\mathbf{y}$, as if the charge density $\rho_j(\mathbf{y})$ of the latter were equal to

$$\rho_j(\mathbf{y}) = u_j^*(\mathbf{y})\,u_j(\mathbf{y}) - \frac{u_j^*(\mathbf{y})\,u_j(\mathbf{x})\,u_i(\mathbf{y})}{u_i(\mathbf{x})} \quad (8.37)$$

instead of only the first term (as in the case of the Hartree equations). With this modification the wave functions become also orthogonal, which is not the case for the solutions of (8.34). One observes from (8.37) that the charge density at $\mathbf{y} = \mathbf{x}$ is zero. A 'hole' of electrons of equal spin is therefore produced about any given electron; it is sometimes called the 'Fermi hole'. However, the total charge seen by any electron remains the same as previously, because the integral over $\mathbf{y}$ of the second term in (8.37) vanishes. Charge is therefore shifted away from the electron, and the repulsive contribution to the energy is reduced. It follows that the Fock equations give lower values for ϵ_i and also for E. We should expect this because the variational principle predicts lower E for better wave functions. Detailed analysis shows in fact that the Fock equations also result from antisymmetrization with respect to *all* valence electrons, (including spin). The Pauli principle between the valence and core electrons is not taken care of, however, and various devices have to be employed in the course of the calculation to enforce it approximately.

The Hartree–Fock method outlined here opens up in principle a way to implement the one-electron picture of Chapter 5 and to test the validity of the results by computing the cohesive energy resulting from the wave functions. That the Hartree–Fock electrons turn out to exist in Bloch orbitals is made plausible from this consideration: without the last term the hamiltonian (8.34) has the lattice periodicity and will yield Bloch functions; if these functions are inserted into the last term they will produce a correction having again lattice periodicity; hence the correction procedure cannot modify $\mathbf{k}$. In principle, therefore, one would think of the calculation for a specific metal running like this:

(*a*) Construction of the core potential, with possible core exchange, covering one primitive cell; the lattice distance is thereby kept as a free parameter.

(*b*) Calculation of the valence electron wave functions in the core field.

(*c*) Computation of the correction to the core potential due to the valence electrons.

(*d*) Alternative repetitions of (*b*) and (*c*) until the self-consistent solutions of (8.34) or its Fock analogue are reached.

(*e*) Computation of the total energy of the crystal as function of the lattice parameter from (8.35), determination of its stable minimum, and comparison with experiment. The experimental energy must be lower than the computed value, because of the variational principle.

Implementation of this programme was started by *Wigner and Seitz*† on sodium. They observed that the part (*b*) of the programme can be formulated entirely within one primitive cell, and if the cell having their name is constructed around an ion core (see figs. 1.30 and 1.31) then the problem (*b*) reduces to solving the wave equation for an almost spherical potential within an almost spherical body. They made use of this aspect to simplify the problem, replacing the Wigner–Seitz cell by a sphere of equal volume. The problem for $\mathbf{k} = 0$ is then spherically symmetric with the condition that there be a vanishing derivative at the surface. Later on, this simplification was coupled with the ex post facto information on alkali metals that electrons and ion cores outside a given cell give a negligible contribution to the potential within that cell; if, furthermore, self-repulsion of the electron cloud within a cell is temporarily neglected, then construction of this wave function can be formulated entirely as a pseudo-atomic programme, namely to find a wave function in the core potential of the atomic ion core which has a vanishing derivative at a prescribed radius (rather than being finite at infinity as for atoms). This divorced problem, thus precisely formulated, is capable of very elegant solution, in such a way that only empirical information from the atom determines the wave function.‡

The elegant procedure (quantum defect method) just outlined applies to the state $\mathbf{k} = 0$ only. For $\mathbf{k} \neq 0$, the boundary conditions create problems to which there are only approximate answers within the Wigner–Seitz framework.§ Herring‖ has therefore proposed an alternative procedure, called the *method of orthogonalized plane waves*. The purpose of the method is to satisfy the boundary conditions and orthogonality requirements first, and thereupon to solve the wave equation. For this purpose a small number of plane waves of the required reduced $\mathbf{k}$ are taken, and the core functions are subtracted from them with suitable multipliers, so as to make them orthogonal to the core. The solution of

---

† E. Wigner and F. Seitz, *Phys. Rev.* **43** (1933), 804; **46** (1934), 509.

‡ T. S. Kuhn and J. H. Van Vleck, *Phys. Rev.* **79** (1950), 382. Frank S. Ham, *Advances in Solid State Physics*, **1** (1955), 127.

§ J. Bardeen, *J. Chem. Phys.* **6** (1938), 367.

‖ C. Herring, *Phys. Rev.* **57** (1940), 1169. C. Herring and A. G. Hill, *Phys. Rev.* **58** (1940), 132.

the wave equation is then reduced to the location of a root of a secular determinant among those functions.

The stages (c) and (d) of the Hartree–Fock programme outlined above are generally believed not to have a strong influence on the electronic wave functions. The reason given is the relative constancy of the wave functions over a great part of the cell, as is evident from fig. 8.14. Self-consistency is therefore enforced only with respect

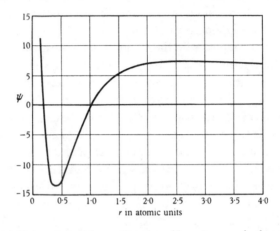

Fig. 8.14. Electronic radial wave function of lowest energy in the valence band of sodium; the wave function is practically a constant in the greater part of the crystal.

to some parameters. One can say therefore that the wave functions employed in this type of work are primarily solutions of a one-particle Schrödinger equation with occasional small modifications based on Hartree–Fock type reasoning.†

The published results on the alkali metals are brought together in Table 4. It appears that metallic binding arises from the 'quantum defect' in the electronic states. In other words, the condition $\partial \psi / \partial n = 0$ at the boundary of the Wigner-Seitz cell represents a relaxed condition in comparison with the boundary condition of the wave function in the free atom. This binding is partly cancelled by the 'Fermi energy': the Pauli exclusion principle forces some

† The only Hartree–Fock type calculation for energy bands known to the author is for aluminium by V. Heine, *Proc. Roy. Soc.* A, **240** (1957), 340.

of the electrons in a pooled system to go into states which are energetically less favourable. The two effects together determine the one-electron energies ϵ_i entering into (8.34). If these quantities are taken with respect to a reference zero such that the Coulomb interaction from neighbouring nuclei is permitted to cancel the electronic part from neighbouring cells (which is the greater part of the second term in (8.32)), then we find that the mean $\bar{\epsilon}$ of these quantities already approximates well the metallic cohesive energy; this is shown in Table 4, columns 4 and 5. Thus if $\bar{\epsilon}$ were the theoretical cohesive energy the calculation would come to a quick successful termination. However, this is not the prediction of the theory. According to (8.32) and (8.35), we must add to this quantity the electron-electron repulsion, which even with our reference zero, is still left for electrons in the same cell. The result is a catastrophe: energies having the wrong sign and three times too large now result. If we pass from (8.35) to the Fock equations, the 'Fermi hole' makes its appearance, which removes electrons of equal spin from each other. It is hard to calculate this contribution to the energy for good crystalline wave functions, but an improvement certainly results. Columns 6 and 7 of Table 4 show the result; we find now roughly zero cohesive energy. Wigner[†] has suggested that the remaining error arises from a correlation in the motion of electrons of opposite spin; for them, too, there is a hole similar to the Fermi hole. Wigner and more recently Gell-Mann and Brueckner[‡] have estimated this effect. The correction gives roughly the right value for the binding energy and lattice constant, as shown in the last two columns of Table 4. It remains true nevertheless, that the one-electron energies alone give already the right answer. To the present theory this appears as an accident for which there is no direct explanation.

The only calculation of this kind for metals which are not alkalis is the work of Herring and Hill[§] on beryllium. Other band calculations have been made, but they are usually broken off before this supreme test.

† E. Wigner, *Phys. Rev.* **46** (1934), 1002. E. Wigner and F. Seitz, *Phys. Rev.* **46** (1934), 509.
‡ M. Gell-Mann and K. A. Brueckner, *Phys. Rev.* **106** (1957), 364.
§ C. Herring and A. G. Hill, *Phys. Rev.* **58** (1940), 132.

Table 4. *Experimental and theoretical information on the alkali metals*

(a in Å and $-E$ in eV per atom.)

Substance	Experimental		One-electron results		Fock results		Corrected results	
	a	$-E$	a	$-E$	a	$-E$	a	$-E$
Li	3·46	1·58	3·8	1·68	3·50	0·63	3·5	1·56
Na	4·25	1·13	3·95	1·25	4·56	0·18	4·14	1·12
K	5·20	0·94	4·50	1·29	5·20	−0·02	4·51	1·21
Rb	5·62	0·89	5·2	1·04	—	—	5·20	1·05

In summary, it appears at this time that we can compute the cohesive energy of solids in principle by quantum mechanics. For solids bonded ionically or by Van der Waals's forces, this possibility of principle can be implemented reasonably well in practice. The character of the covalent and metallic bond, on the other hand, is such as to make each substance a separate problem. The bonding is provided in this case by the accumulation of electrons between the ion cores. This accumulation is governed by Schrödinger's equation; it shifts from case to case and is difficult to calculate for mechanical and conceptual reasons. The general theoretical possibility will therefore be implemented only for a small number of materials, as long as the theory retains its present cumbersome form. Generalizations which are both simple and precise might conceivably alter this prospect.

PROBLEMS

1. The computation of the Madelung number demands the computation at the site of one ion of the potential due to all other ions. Evjen showed that the calculation can be carried out in this simple form with rapid convergence if the ions surrounding the singled out ion are grouped into cubically symmetric shells of no resultant charge; to achieve this some charges may have to be split into fractions. Outline a plan for such a calculation to obtain the numbers (8.5), and carry the method into execution.

2. Derive the expression (8.37) for the effective charge density with exchange for two electrons of parallel spin, starting with the hamiltonian (8.29) and the wave function (8.36).

3. Show that if we employ a completely antisymmetrized wave function, such as the wave function (5.85) rather than (8.36) we still end up with the same Fock equations.

4. Introduce a magnetic field into the hamiltonian (8.29) by the substitutional device

$$p_i \rightarrow p_i - \frac{e}{c} A(x_i)$$

and show that now the variational principle (8.27) leads to the refinement (8.33).

BIBLIOGRAPHY

REFERENCE MATERIAL FROM ATOMIC THEORY

C. E. MOORE. *Atomic Energy Levels.* U.S. National Bureau of Standards Circular 467 (Washington 1949 and 1952). 2 vols.

D. R. HARTREE. The Calculation of Atomic Structures. (New York, Wiley, 1957.)

W. DÖRING. Elektronenverteilung in Atomen und Ionen nach Hartree. Landolt–Börnstein, *Zahlenwerte und Funktionen aus Physik, Chemie, Astronomie, Geophysik und Technik,* Pt. 1, pp. 276–96 (1950).

R. S. KNOX. Bibliography of atomic wave functions. *Solid State Physics,* **4** (1957), 413–21.

THEORY OF VAN DER WAALS FORCES

H. MARGENAU. Van der Waals forces. *Rev. Mod. Phys.* **11** (1939), 1–35.

THEORY OF MOLECULAR STRUCTURE

L. PAULING and E. B. WILSON. *Introduction to Quantum Mechanics.* (New York, McGraw-Hill, 1935.)

H. A. BETHE and E. E. SALPETER. Quantum mechanics of one- and two-electron systems. *Handbuch der Physik,* **35** (1957), 88–436.

H. HARTMANN. Theorie der chemischen Bindung auf quantentheoretischer Grundlage. (Berlin, Springer, 1954.)

Y. K. SYRKIN and M. E. DIATKINA. *Structure of Molecules and the Chemical Bond* (translation). (New York, Interscience, 1950.)

L. PAULING. *Nature of the Chemical Bond and the Structure of Molecules and Crystals.* 2nd ed. (Ithaca, Cornell Univ. Press, 1945.)

G. HERZBERG. *Molecular Spectra and Molecule Structure.* 2nd ed. (New York, Van Nostrand, 1950.) 2 vols.

A. G. GAYDON. *Dissociation Energies and Spectra of Diatomic Molecules.* (New York, Wiley, 1947.)

T. L. COTTRELL. *The Strengths of Chemical Bonds.* (London, Butterworth, 1954.)

F. H. FIELD and J. L. FRANKLIN. *Electron Impact Phenomena and the Properties of Gaseous Ions.* (New York, Academic Press, 1957.)

F. R. BICHOWSKY and F. D. ROSSINI. *The Thermochemistry of the Chemical Substances.* (New York, Reinhold, 1936.)

STRUCTURE OF IONIC CRYSTALS

J. SHERMAN. Crystal energies of ionic compounds and thermochemical applications. *Chem. Rev.* **11** (1932), 93–170.

M. BORN and M. GÖPPERT-MAYER. Dynamische Gittertheorie der Kristalle. *Handb. d. Phys.* **24**, part 2 (1933), 623–794.

M. BORN and K. HUANG. *Dynamical Theory of Crystal Lattices.* (Oxford, Clarendon Press, 1954).

THEORY OF METALLIC COHESION

F. SEITZ. *Modern theory of solids*, chapter 10. (New York, McGraw-Hill, 1940.)

J. R. REITZ. Methods of the one-electron theory of solids. *Solid State Physics*, **1** (1955), 1–95.

E. P. WIGNER and F. SEITZ. Qualitative analysis of the cohesion in metals. *Solid State Physics*, **1** (1955), 97–126.

F. S. HAM. The quantum defect method. *Solid State Physics*, **1** (1955), 127–92.

T. O. WOODRUFF. The orthogonalized plane wave method. *Solid State Physics*, **4** (1957), 367–411.

INDEX

Phonons, *see* Lattice vibrations, *or*
Polarization of phonons, *or* Absorption
of phonons, *or* Emission of
phonons
Photons, absorption of, *see* Absorption
of photons
Piezoelectricity, 129
Pinsker, Z. G., 49
Plane of symmetry, 8–10, **14–16**, 33, 237
Point group, 2
Polarization, electric, 121, 127
Polarization of phonons, 64–68, 79–82,
220–223
Polarized light, 32
Polaron, 169
Potassium, 256
Potassium bromide, 228, 232
Potassium chloride, 228, 232
Potassium dihydrogen phosphate, 120,
121
Potassium fluoride, 228
Pottassium iodide, 228, 232
Primitive cell of crystals, **5**, **6**, 15–22,
36, 148, 151, 171, 253
Primitive cell of reciprocal lattice, 48,
74, 86
Primitive cell as scattering unit, 40, 43,
64
Principle of spectroscopic stability,
217
Pyroelectric crystals, 120

Quantization, 55, 76, 78, 90, 99, 190,
196, 224
Quantum defect method, 253, 254,
258
Quantum mechanics, *see* Schrödinger
equation or Matrix mechanics, *or*
Matrix element, *or* Quantum theory
Quantum theory, 76, 82–85, 99, 190,
200–205, 212, 232, 236, 248, 256
Quartz, 3, 31–33, 89, 117–121
Quasi-classical band theory, 190–199,
203, 206
Quasi-wave function, *see* Cellular
amplitude function

Radii of ions, 229
Raman scattering of light, 82–84
Raman scattering of matter waves,
83–86, 219
Rare earths, 163
Rare gases, *see* Noble gases
Reciprocal lattice, **44–48**, 65, 83–85,
176

Reciprocal mass tensor *see* Mass,
tensor, reciprocal
Reciprocal space, 44, 45, 48, 197, *see
also* Wave number space
Rectangular Bravais net, 10, 16
Reduced representation of groups,
60–63
Reduced wave vector, 47, 53, 56, 142,
145
Reflection symmetry, *see* Plane of
symmetry
Reitz, J. R., 164, 257
Relation of energy versus wave vector,
see Energy versus wave vector
relation
Relativity theory, 194
Relaxation mechanism, 206–209, 214,
217, **219–223**
Relaxation of density matrix, **217**,
220–222
Relaxation rate matrix, 217
Relaxation theory for electron scattering,
205–217, 219
Relaxation time, 207–225
Remanence, 94, 95
Representation of group, 59–63, 86,
87
Representation module, 60–63
Resistivity, *see* Conductivity
Resonance, paramagnetic, *see* Paramagnetic
resonance
Resonance stabilization, 238
Reststrahl band, 59
Rhombic Bravais, net, 10, 16
Rhombohedral crystals, **18–21**, **28**, **29**,
36, 37, 48
Rice, O. K., 126
Richardson law, 158
Rigid cell model, 220
Rocksalt, *see* Sodium chloride
Rossini, F. D., 257
Rotational symmetry, *see* Axis of
symmetry
Rubidium, 256
Rubidium bromide, 232
Rubidium chloride, 232
Rubdidium iodide, 232
Rundle, R. E. 246
Rutherford's law of scattering, 219

σ-bond, 245
Salpeter, E. E., 257
Salts, *see* Ionic crystals
Saturated crystals, *see* Chemical saturation